STUDENT'S STUDY GUIDE AND SOLUTIONS MANUAL

JAMES LAPP

USING & UNDERSTANDING MATHEMATICS: A QUANTITATIVE REASONING APPROACH

SIXTH EDITION

Jeffrey Bennett

University of Colorado at Boulder

William Briggs

University of Colorado at Denver

PEARSON

Boston   Columbus   Indianapolis   New York   San Francisco   Upper Saddle River
Amsterdam   Cape Town   Dubai   London   Madrid   Milan   Munich   Paris   Montreal   Toronto
Delhi   Mexico City   São Paulo   Sydney   Hong Kong   Seoul   Singapore   Taipei   Tokyo

The author and publisher of this book have used their best efforts in preparing this book. These efforts include the development, research, and testing of the theories and programs to determine their effectiveness. The author and publisher make no warranty of any kind, expressed or implied, with regard to these programs or the documentation contained in this book. The author and publisher shall not be liable in any event for incidental or consequential damages in connection with, or arising out of, the furnishing, performance, or use of these programs.

Reproduced by Pearson from electronic files supplied by the author.

ISBN-13: 978-0-321-91532-0
ISBN-10: 0-321-91532-1

1 2 3 4 5 6 EBM 18 17 16 15 14

www.pearsonhighered.com

PEARSON

Table of Contents

Study Guide

Solutions

Introduction

Welcome to the *Student's Study Guide and Solutions Manual* for *Using and Understanding Mathematics: A Quantitative Reasoning Approach, Sixth Edition.* Hopefully, with the help of this book, your course in quantitative reasoning will be both enjoyable and successful. The goal of this guide is not to add to your workload, but to give you a set of concise notes that will make your studying as effective as possible. If you work with this guide as you read and do exercises, you should get the most from the course.

This guide is organized according to the units in the textbook. For each unit you will find the following features:

Overview

This section provides a brief survey of the unit, its major points and its goals. This feature is not a substitute for reading the text.

Key Words and Phrases

This section is simply a list of the important words and phrases used in the unit. You can use this section for review, study, and self-testing. You should be able to explain or define all of the terms on this list.

Key Concepts and Skills

In this section you will find a summary of the most important concepts in each unit. These concepts may be general ideas (for example, the distinction between deductive and inductive arguments) or basic skills (for example, creating the equation of a straight line). This section should also be helpful for review, study, and self-testing.

Important Review Boxes

If a unit has one or more Review Boxes, they are listed. These boxes are quite important for providing background skills or knowledge needed for the unit.

Solutions to (Most) Odd Problems

Following all of the unit summaries, the second half of this Guide provides full solutions to most of the odd-numbered problems in the textbook. Mathematics is not a spectator sport. Reading the solutions is never a substitute for working the problems. *You are strongly advised to work the problems first and then check the solutions.*

How to Succeed In This Course

Using *This* Book

Before we get into more general strategies for studying, here are a few guidelines that will help you use *this* book most effectively.

- Before doing any assigned problems, read assigned material *twice*:
  On the first pass, read quickly to gain a feel for the material and concepts presented.
  On the second pass, read the material in more depth, and work through the examples carefully.
- During the second reading, take notes that will help you when you go back to study later. In particular, *use the margins*. The wide margins in this textbook are designed to give you plenty of room for making notes as you study.
- After you complete the reading, and again when studying for exams, make sure you can answer the *review questions* at the end of each unit.
- You'll learn best by *doing*, so work plenty of the end-of-unit exercises. Don't be reluctant to work more than the exercises that your instructor assigns.

Budgeting Your Time

A general rule of thumb for college courses is that you should expect to study about 2 to 3 hours per week *outside* class for each unit of credit. For example, a student taking 15 credit hours should spend 30 to 45 hours each week studying outside of class. Combined with time in class, this works out to a total of 45 to 60 hours per week — not much more than the time required of a typical job. If you find that you are spending fewer hours than these guidelines suggest, you can probably improve your grade by studying more. If you are spending more hours than these guidelines suggest, you may be studying inefficiently; in that case, you should talk to your instructor about how to study more effectively for a mathematics class.

General Strategies for Studying

- Don't miss class. Listening to lectures and participating in discussions is much more effective than reading someone else's notes. Active participation will help you retain what you are learning.
- Budget your time effectively. An hour or two each day is more effective, and far less painful, than studying all night before homework is due or before exams.
- If a concept gives you trouble, do additional reading or problem solving beyond what has been assigned. If you still have trouble, *ask for help*: you surely can find friends, colleagues, or teachers who will be glad to help you learn. Never be reluctant to ask questions or ask for help in this course. If you have a question or problem, it is extremely unlikely that you will be alone.
- Working together with friends can be valuable; you improve your own understanding when discussing concepts with others. However, be sure that you learn *with* your friends and do not become dependent on them.

Preparing for Exams

- Rework exercises and other assignments; try additional exercises to be sure you understand the concepts. Study your assignments, quizzes, and exams from earlier in the semester.
- Study your notes from lectures and discussions. Pay attention to what your instructor expects you to know for an exam.
- Reread the relevant sections in the textbook, paying special attention to notes you have made in the margins.
- Study individually before joining a study group with friends. Study groups are effective only if *every* individual comes prepared to contribute.
- Try to relax before and during the exam. If you have studied effectively, you are capable of doing well. Staying relaxed will help you think clearly.

Finally, good luck. We wish you an enjoyable and rewarding experience with quantitative reasoning.

1 Thinking Critically

Overview

Before discussing Chapter 1, we urge you to take a few minutes to read the prologue to the textbook. This short chapter sets the stage for the entire book. It presents the idea of quantitative reasoning and discusses the importance of interdisciplinary thinking. It gives a high altitude picture of mathematics and how it impacts many other subjects that you will encounter either in other courses or in your career. Finally, it gives some advice on using the book and studying for your course. It's worth a quick reading.

In teaching this course to many students over many years, we know that often the most serious weakness that students bring to the course is not poor mathematical skills, but poor reasoning skills. It's not multiplying two numbers that creates problems, but deciding *when* to multiply. For this reason, the book opens with a chapter that contains virtually no mathematics (in the sense of manipulating equations and finding their solutions). The emphasis of the chapter is critical thinking and logical skills.

In this chapter you will encounter some introductory logic, but don't worry; we don't get carried away with symbolic logic and heavy-duty truth tables. In fact, much of this chapter may be familiar to you from a previous course in logic or philosophy.

Unit 1A Living in the Media Age

Unit 1A opens the chapter by explaining that a logical **argument** (as opposed to an everyday argument) is a set of facts or assumptions, called **premises**, that lead to a **conclusion**. A **fallacy** is an argument that is either deceptive or wrong. This unit explores common fallacies that you might encounter in advertising or news reports. We present ten informal fallacies, some of which may seem obvious, others of which may be quite subtle. Critical reading and thinking will help you avoid becoming a victim of these fallacies.

Key Words and Phrases

logic	argument	premise
conclusion	fallacy	appeal to popularity
false cause	appeal to ignorance	hasty generalization
limited choice	appeal to emotion	personal attack
circular reasoning	Diversion (Red Herring)	straw man

Key Concepts and Skills

- Identify the premise and conclusion of an argument.
- Recognize informal fallacies in advertisements and news reports.
- Five Steps to Evaluating Media Information: 1) Consider the source. 2) Check the date. 3) Validate accuracy. 4) Watch for hidden agendas. 5) Don't miss the big picture.

Unit 1B Propositions and Truth Values

Overview

In this unit formal logic is introduced in a somewhat casual manner. We start with **propositions** — statements that make a claim that can be true or false. Then we look at the **connectors** that can be used with propositions to make more complex propositions. The connectors that you will encounter are

- *not* (negation)
- *or* (disjunction)
- *and* (conjunction)
- *if ... then* (implications).

Whereas many logic books make heavy use of symbolic logic and truth tables, we will use truth tables primarily for fairly simple propositions that involve one, two, or three connectors. So our excursion into symbolic logic will be limited and designed to provide only an introductory glimpse.

The *if ... then* connector is quite important in both logic and everyday speech (for example, *if I pass this course, then I will graduate*). For this reason, we spend a little time discussing other forms of the proposition *if p, then q*. These other forms are the

- **converse** (*if q, then p*)
- **inverse** (*if not p, then not q*) and
- **contrapositive** (*if not q, then not p*).

This particular discussion may seem a bit technical, but it's also extremely practical. For example, suppose it's true that *if I read the book, then I will pass the course*. Does it follow that *if I don't read the book, then I won't pass the course*?

Key Words and Phrases

proposition	truth table	negation
conjunction	disjunction	conditional
antecedent	consequent	converse
inverse	contrapositive	logical equivalence

Key Concepts and Skills

- Understand negation, conjunction, disjunction, and conditionals, and their truth tables.
- Use truth tables to evaluate the truth of compound propositions that use two or more connectors.
- Analyze various forms of *if ... then* propositions.

Unit 1C Sets and Venn Diagrams

Overview

You may have encountered Venn diagrams before now as a way to illustrate the relationships between collections of objects, or **sets**. In this unit, we review some of the most basic properties of sets and then discuss how Venn diagrams can be used to work with sets. Throughout the unit, the emphasis is on practical applications of sets and Venn diagrams. One of the important applications of Venn diagrams is to illustrate what are called **categorical propositions** of logic. We will study four basic categorical propositions. Given a **subject set** S and a **predicate set** P, they may be related in the following ways:

- All S are P (for example, all whales are mammals)
- No S are P (for example, no fish is a mammal)
- Some S are P (for example, some doctors are women)
- Some S are not P (for example, some teachers are not men).

As we will see, each form of categorical proposition has a specific Venn diagram. Equally important, the negation of each categorical proposition is one of the other propositions in the list above. Specifically, we have the following relations between the four propositions and their negations.

Proposition	Negation
All S are P	Some S are not P
No S are P	Some S are P
Some S are P	No S are P
Some S are not P	All S are P

The unit concludes with a discussion of the additional uses of Venn diagrams.

Key Words and Phrases

set	Venn diagram	subset
disjoint sets	overlapping sets	categorical propositions

Key Concepts and Skills

- Use set notation.
- Construct Venn diagrams for categorical propositions.
- Put propositions in standard form.
- Construct Venn diagrams for three or more sets.

Important Review Box

- Brief Review: Sets of Numbers

Unit 1D Analyzing Arguments

Overview

The propositions that we studied in the previous unit can be combined in various ways to form arguments. Of primary importance is the distinction between **deductive** and **inductive** arguments. Deductive arguments generally proceed from general premises to a more specific conclusion. In a deductive argument, all of the premises are needed to reach the conclusion. By contrast, inductive arguments generally proceed from specific premises to a general conclusion; in an inductive argument the premises independently support the conclusion.

To analyze inductive arguments, we ask about the **strength** or **weakness** of the argument. Determining the strength of an inductive argument is often a subjective judgment, and so there are no systematic methods to apply.

Most of the unit is spent analyzing three-line deductive arguments using Venn diagrams. They can consist of the four types of categorical propositions or they may involve conditional propositions (studied in Unit 1C). Of fundamental importance in this business is the distinction between valid and invalid arguments. An argument is **valid** if, based on the Venn diagram analysis, it is logically solid and consistent. An argument that fails the Venn diagram analysis must contain a fallacy and is **invalid**. Validity has nothing to do with the truth of the premises or conclusion, rather, it is a measure of the logical structure of the argument.

Having shown that an argument is valid, we can then ask if it is sound. A **sound** argument is valid *and* has true premises; a sound argument must lead to a true conclusion. Soundness is the highest test of a deductive argument.

Fallacies can arise in deductive arguments in many ways. Perhaps the most common fallacies occur in arguments that involve conditional (*if...then*) propositions. These fallacies appear in the everyday arguments of advertising and news reports. There are four different forms of conditional arguments; two are valid and two are invalid:

- affirming the hypothesis (valid)
- affirming the conclusion (invalid)
- denying the hypothesis (invalid)
- denying the conclusion (valid).

The unit closes with examples of how inductive and deductive arguments are used in mathematics.

Key Words and Phrases

deductive	inductive	strength/weakness
valid/invalid	sound	affirming the hypothesis
affirming the conclusion	denying the hypothesis	denying the conclusion
Pythagorean theorem		

Key Concepts and Skills

- Know the distinction between deductive and inductive arguments.
- Determine the strength of inductive arguments.
- Assess the validity of three-line deductive arguments consisting of categorical propositions, using Venn diagrams.
- Assess the validity of three-line deductive arguments involving conditional propositions, using Venn diagrams.
- Identify fallacies that arise in conditional arguments.
- Understand various combinations of valid/invalid and sound/unsound that can occur in deductive arguments.
- Determine the soundness of three-line deductive arguments.

Unit 1E Critical Thinking in Everyday Life

Overview

Critical thinking is an approach to problem solving and decision-making that involves careful reading (or listening), sharp thinking, logical analysis, good visualization, and healthy skepticism. In this chapter, we present several guidelines designed to sharpen critical thinking skills as they apply to practical problems:

- Read or listen carefully.
- Look for hidden assumptions.
- Identify the real issue.
- Understand all the options.
- Watch for fine print and misinformation.
- Are other conclusions possible?
- Don't miss the big picture.

Key Concepts and Skills

- Apply the guidelines of the unit to practical decisions and problems.

2   Approaches to Problem Solving

Overview

The first chapter of the book was devoted to *qualitative* issues — topics that don't require extensive use of numbers and computation. In this chapter (and the remainder of the book) we turn to *quantitative* matters. Perhaps it's not surprising that we begin our study of quantitative topics with problem solving. The first two units of the chapter deal with a very basic and important problem-solving technique, the use of units. The last unit of the chapter presents various problem-solving strategies. It is a valuable chapter whose lessons run through the rest of the book.

Unit 2A Working with Units

Overview

Nearly every number that you encounter in the real world is a measure of *something*: 6 billion *people*, 5280 *feet*, 9 trillion *dollars*, 26 *cubic feet*. The quantities that go with numbers are called **units**. The message of this section of the book is that using units can simplify problem solving immensely. Indeed the use of units is one of the most basic problem solving tools.

We begin by considering the most basic **simple units** for the fundamental types of measurement. Here are a few examples of simple units.
- length — inches, feet, meters.
- weight — pounds, grams.
- volume — quarts, gallons, liters.
- time — seconds, hours.

From these simple units, we can build endless **compound units**.

The rest of the unit is spent illustrating a wide variety of such compound units. Among the many you will meet and work with are…
- units of area such as square feet and square yards.
- units of volume such as cubic inches and cubic feet.
- units of speed such as miles per hour.
- units of price such as dollars per pound.
- units of gas mileage such as miles per gallon.

One of the realities of life is that there are many units for the same quantity. For example, we can measure lengths in inches, centimeters, feet, meters, miles, or kilometers. We measure floor area in square feet, but have to buy carpet in square yards. Thus one of the necessities of problem solving is being able to convert from one unit to another consistent unit. The key to carrying out conversions between units is to realize that there are three equivalent ways to express a conversion factor. For example, we can say 1 foot = 12 inches, or we can say 1 foot *per* 12 inches, or we can say 12 inches *per* foot. Mathematically, we can write

$$1 \text{ ft} = 12 \text{ in. or } \frac{1 \text{ ft}}{12 \text{ in.}} = 1 \text{ or } \frac{12 \text{ in.}}{1 \text{ ft}} = 1.$$

These three forms of the same conversion factor are equivalent. The key to a happy life with units is choosing the appropriate form of the conversion factor for a given situation.

Additional problem-solving skills can then be built on these fundamental skills. You will see and learn how to make a chain of conversions factors to solve more complex problems. For example, the time required to count a billion dollars at the rate of one dollar per second is

$$\$1,000,000,000 \times \frac{1 \text{ sec}}{\$1} \times \frac{1 \text{ min}}{60 \text{ sec}} \times \frac{1 \text{ hr}}{60 \text{ min}} \times \frac{1 \text{ day}}{24 \text{ hr}} \times \frac{1 \text{ yr}}{365 \text{ days}} = 31.7 \text{ years.}$$

The fact that the units cancel and give an answer in *years* tells you that the problem has been set up correctly.

We expand the study of units and conversion factors by examining two standard systems of units: The U.S. Customary System of Measurement (or USCS system, which is used primarily in the United States) and the metric system (which is used everywhere else in the world).

We first proceed systematically and survey the USCS units for length, weight, and volume. Tables 2.3 and 2.4 contain many conversion factors, but you should focus on *using* these conversion factors, not memorizing them. Having seen the complications of the USCS system, the metric system should come as a welcome relief. Next we present the metric units for length, weight and volume. The system is based on powers of ten and standard prefixes (Table 2.2), which makes conversions between units relatively simple.

Finally, we look at a very practical type of unit conversion problem, those associated with currency. If you travel to Mexico you might find 1 peso is equal to about 7 cents. From this fact, you might need to answer questions such as:

- Which is larger, 1 peso or 1 dollar?
- How many pesos are in a dollar?
- How many dollars are in a peso?
- How many dollars are in 23.45 pesos?
- If apples cost 23 pesos per kilogram, what is the price in dollars per pound?

Here are a few final words of advice: Never was the motto *practice makes perfect* more true than with problem solving and units. You should work all assigned exercises, and then some, in order to master these techniques. And unless your instructor tells you otherwise, there is no need to memorize hundreds of conversion factors. It's helpful to know a few essential conversion factors off the top of your head. As for the rest, it's easiest just to know how to find them quickly in the book.

Key Words and Phrases

simple units	compound units	area
volume	conversion factor	USCS system
metric system	meter	gram
liter	second	

Key Concepts and Skills

- Convert from one simple unit to another; for example, from inches to yards.
- Convert from one unit of area to another unit of area; for example, from square inches to square yards.
- Convert from one unit of volume to another unit of volume; for example, from cubic inches to cubic yards.
- Solve problems involving chains of conversion factors; for example, finding the number of seconds in a year.
- Convert from one unit of currency to another; for example, from dollars to pesos.
- Given the required conversion factor, convert between two consistent USCS units; for example, rods to miles.
- Multiply and divide powers of ten.
- Know basic metric units and commonly used prefixes.
- Given the required conversion factor, convert between two consistent metric units; for example, millimeters to kilometers.
- Given the required conversion factor, convert between a USCS unit and a consistent metric unit; for example, cubic inches to liters.
- Convert between temperatures in the three standard systems.
- Solve problems using chains of conversion factors involving USCS units, metric units, and currency units.

Important Review Box

- Brief Review: Common Fractions
- Brief Review: Decimal Fractions
- Brief Review: Powers of 10

Unit 2B Problem Solving with Units

Overview

Units can aid in problem solving. Their use can be summarized by the following:

Unit Analysis in Problem Solving

- Step 1. Identify the units involved in the problem and the units that you expect for the answer.
- Step 2. Use the given units and the expected answer units to help you find a strategy for solving the problem.
- Step 3. When you complete your calculations, make sure that your answer has the units you expected

In this unit, we also explore two more categories of units. **Energy** (what makes things move or heat up) and **power** (the rate at which energy is used) are incredibly important concepts in understanding the world around us. Uses of energy and power units are presented in matters such as utility bills, diet, and the environment. The units of **density** and **concentration** are also discussed in detail. In measuring population density, the capacity of computer discs, levels of pollution, and blood alcohol content, these units are indispensable.

Armed with all of these units and conversion factors (and don't forget currency conversion factors as well), we can do even more elaborate problem solving. This is the goal of the remainder of the unit and the exercises at the end of the unit.

Key Words and Phrases

Kelvin	energy	power
kilowatt-hour	watt	density
concentration	population density	

Key Concepts and Skills

- Understand how units can help in problem solving.
- Understand and apply units of energy and power.
- Understand and apply units of density to materials, population, and information.
- Understand and apply units of concentration to pollution, and blood alcohol content.

Unit 2C Problem-Solving Guidelines and Hints

Overview

Having seen very specific examples of problem solving in the previous two units, we now present some strategies for problem solving in general. There is no simple and universal formula for solving all problems. Only continual (and hopefully enjoyable) practice can lead one towards mastery in problem solving. This unit is designed to provide you some of that practice.

The unit opens with a very well known four-step process for approaching problem solving. This process is not a magic formula, but rather a set of guidelines. The four basic steps are:
- Understand the problem,
- Devise a strategy,
- Carry out the strategy, and
- Look back, check, interpret and explain your solution.

The remainder of the unit is a list of eight strategic hints for problem solving. Here are the strategic hints:
- There may be more than one answer.
- There may be more than one strategy.
- Use appropriate tools.
- Consider simpler, similar problems.
- Consider equivalent problems with simpler solutions.
- Approximations can be useful.
- Try alternative patterns of thought.

• don't spin your wheels.

Most of the unit consists of examples in which the four-step process and these strategic hints are put to use. Study these problems and solutions, and try to see how the techniques might be used in other problems that you encounter.

Key Concepts and Skills

• Carry out the four-step problem solving process.
• Identify and use the eight strategic problem solving hints.

3   Numbers in the Real World

Overview

In this chapter we explore how numbers are used in real and relevant problems. The first unit deals with percentages and is arguably one of the most practical and important units of the book. As you know, many real world numbers are incredibly large (the federal debt or the storage capacity of a computer disk) or very small (the diameter of a cancer cell or the wavelength of an x-ray), so in the next unit, we introduce scientific notation to deal with large and small numbers. Unit 3C discusses another reality of numbers in the real world: they are often approximate or subject to errors. The news is filled with reference to index numbers; for example, the consumer price index, or CPI. Such index numbers are critical to an understanding of the news, so they are the subject of Unit 3D. Finally, the last unit is a fascinating exploration of how numbers deceive us. All in all, it's a very practical and useful chapter.

Unit 3A Uses and Abuses of Percentages

Overview

If you read a news article, a financial statement, or an economic report, you may notice that one of the most common ways to communicate quantitative information is with percentages. In this unit, we will explore the many ways that percentages are used — and abused. Percentages are used for three basic purposes:

- as fractions (for example, 45% of the voters favored the incumbent),
- to describe change (for example, taxes increased by 10%), and
- for comparison (for example, women live 3% longer than men).

In this unit, we examine each of these uses of percentages in considerable detail with plenty of examples taken from the news and from real situations. The use of percentages as fractions is probably familiar. The use of percentages to describe change relies on the notions of **absolute change** and **relative** (or **percentage**) **change**. When some quantity changes from a previous value to a new value, we can define its change in either of two ways:

$$\text{absolute change} = \text{new value} - \text{reference value}$$

$$\text{relative change} = \frac{\text{absolute change}}{\text{reference value}} = \frac{\text{new value} - \text{reference value}}{\text{reference value}}.$$

The use of percentages for comparison relies on the notions of **absolute difference** and **relative** (or **percentage**) **difference**. To make comparisons between two quantities, we must identify the **compared quantity** (the quantity that we are comparing) and the **reference quantity** (the quantity to which we make the comparison). For example, if we ask how much larger is one meter than one yard, *one meter* is the compared quantity and *one yard* is the reference quantity. Then we can calculate two kinds of difference:

$$\text{absolute difference} = \text{compared value} - \text{reference value}$$

$$\text{relative difference} = \frac{\text{absolute difference}}{\text{reference value}} = \frac{\text{compared value} - \text{reference value}}{\text{reference value}}.$$

A powerful rule for interpreting statements involving percentages is what we call the **of versus more than rule**. It says that if the compared value is *P% more than* the reference value, then it is $(100 + P)\%$ *of* the reference value, and if the compared value is *P% less than* the reference value, it is $(100 - P)\%$ *of* the reference value. For example, if Bess' salary is 30% more than Bob's salary, then Bess' salary is 130% *of* Bob's salary. Similarly, if Jill's height is 30% *less than* Jack's height, then Jill's height is 70% *of* Jack's height.

All of these ideas are assembled in this unit to solve a variety of practical problems. Our experience in teaching this subject is that the greatest difficulty is understanding the problem and translating it into mathematical terms. It is important to read the problem carefully, draw a picture if necessary, decide how percentages are used in the problem (as a fraction, for change, or for comparison), and to write a mathematical sentence that describes the situation. It is important to study the examples in the unit and work plenty of practice problems.

The unit closes with several examples of ways that percentages are abused. Be sure you can identify problems in which the previous value shifts and problems in which the quantity of interest is itself a percentage.

Key Words and Phrases

absolute change	relative change	percentage change
reference quantity	compared quantity	absolute difference
relative difference	percentage difference	*of* versus *more than* rule

Key Concepts and Skills

- Identify and solve problems that use percentages as fractions.
- Identify and solve problems that use percentages to describe change.
- Identify and solve problems that use percentages for comparison.
- Apply the *of* versus *more than* rule to practical problems.
- Identify fallacies and errors in statements involving percentages.

Important Review Boxes

- Brief Review: Percentages
- Brief Review: What Is a Ratio?

Unit 3B Putting Numbers in Perspective

Overview

Numbers in the world around us are often very large or very small. In order to write large and small numbers compactly, without using a lot of zeros, we use **scientific notation**. This is the main new mathematical idea in this unit. If you haven't used scientific notation before, you will want to practice writing numbers in scientific notation, and multiplying and dividing numbers in scientific notation.

We next spend some time showing how to make rough calculations using estimation. Often we can use approximate values of quantities and arrive at useful **order of magnitude** answers.

All of us suffer from number numbness: large and small numbers eventually lose all meaning. For example, most of us have no sense of how large $12 trillion (the federal debt) or 7 billion (the world's population) really are. Therefore, just as important as *writing* large and small numbers is *visualizing* large and small numbers. We look at some fascinating methods, such as **scaling**, for giving large and small numbers meaning. The goal is often to associate numbers with a striking visual image. For example, if the earth is the size of a ball-point pen tip, the sun would be a grapefruit 15 meters away.

Key Words and Phrases

scientific notation	order of magnitude	scaling
scale ratio	light-year	

Key Concepts and Skills

- Write numbers in scientific notation.
- Multiply and divide numbers in scientific notation.
- Use estimates to compute order of magnitude answers.
- Use scaling methods for visualizing large and small numbers.

Important Review Boxes

- Brief Review: Working with Scientific Notation

Unit 3C Dealing With Uncertainty

Overview

While they may appear totally reliable, the numbers we see in the news or in reports often are quite uncertain and prone to errors. This unit reminds us of this fact in many different ways, as we consider the sources of errors and the measurement of uncertainty in numbers.

The precision of a number can be described by the number of **significant digits** it has. Significant digits are those digits in a number that we can assume to be reliable, although, this is where care must be used in reading numbers. It's usually easy to determine the number of significant digits in a number; there are a few subtle cases that are listed in the summary box in the unit.

Errors enter problems and calculation in two ways. Errors that arise in unpredictable and unavoidable ways are called **random errors**. On the other hand, **systematic errors** arise due to a problem in the system that affects all measurements in the same way; these errors can often be avoided.

We next discuss **absolute errors** and **relative errors**. It is important to note the analogy between absolute/relative errors and absolute/relative difference (discussed in Unit 3A).

The next observation in the unit concerns the difference between **accuracy** and **precision**. A pharmacist's scale that can weigh fractions of an ounce has much more *precision* than a bathroom scale that can measure only in pounds. Thus precision refers to how precisely a quantity can be measured or how precisely a number is reported.

On the other hand, the *accuracy* of a measurement refers to how close it is to the true value (which we often don't know). Suppose a marble weighs 4.5 ounces. If one scale reports a weight of 4.4 ounces and another scale reports a weight of 4.7 ounces, the first measurement is more accurate; both measurements have the same precision — to the nearest tenth of an ounce.

A few technicalities arise when it comes to doing arithmetic with approximate numbers. There are two rules that tell us how much precision should be assigned to the sum, difference, product, or quotient of two approximate numbers.
- when adding or subtracting two approximate numbers, the result should be rounded to the same precision as the *least precise* number in the problem.
- when multiplying or dividing two approximate numbers, the result should be rounded to the same number of significant digits as the number in the problem with the *fewest significant digits*.

Key Words and Phrases

significant digits	random errors	systematic errors
absolute error	relative error	accuracy
precision		

Key Concepts and Skills
- Determine the number of significant digits in a given number.
- Distinguish between random errors and systematic errors.
- Compute absolute and relative errors.
- Understand the difference between accuracy and precision.
- Determine the precision of the sum or difference of two approximate numbers.
- Determine the precision of the product or quotient of two approximate numbers.

Important Review Boxes
- Brief Review: Rounding

Unit 3D Index Numbers: The CPI and Beyond

Overview

Suppose you want to compare the price of gasoline today with its price 25 years ago. Simply comparing the current price with the old price does not give a realistic comparison, because the price of just about everything increased during those 25 years. In this unit, we study the use of index numbers to make such comparisons more realistic. Among the most familiar index number is the **Consumer Price Index** (CPI), which essentially measures the cost of living and its changes due to **inflation**. There are many other index numbers, some of which will be explored in this unit, and all can be summarized by the following rule.

An **index number** provides a simple way to compare measurements made at different times or in different places. A value at one particular time (or place) is chosen as the *reference value*, and the *index number* for any other time (or place) is given by

$$\text{index number} = \frac{\text{value}}{\text{reference value}} \times 100.$$

Prices can be adjusted for inflation by applying the following principle: given a price in dollars for year X ($\$_x$), the equivalent price in dollars for year Y ($\$_Y$) is:

$$\text{price in } \$_Y = (\text{price in } \$_X) \times \frac{\text{CPI}_Y}{\text{CPI}_X}$$

where X and Y represent years, such as 1992 and 2007.

Key Words and Phrases

index number	reference value	Consumer Price Index
inflation	rate of inflation	

Key Concepts and Skills

- Compute index numbers and use them to make comparisons.
- Use the CPI to compare prices.
- Adjust prices for inflation.

Unit 3E How Numbers Can Deceive: Polygraphs, Mammograms, and More

Overview

The topics in this unit may seem a bit unrelated, but they are connected by the common theme of how numbers can be deceptive. The unit begins with a curious phenomenon known as **Simpson's paradox**. It can occur in many different ways, but it always arises when quantities are averaged and some crucial information is missing. The text examples illustrate the effect in many ways and should be studied carefully.

We next turn to surprisingly deceptive results that occur when using percentages in what are called *two-by-two tables* (such as Table 3.7). These tables are used to describe two different outcomes (for example, cured and not cured) for two different groups of people (for example, treatment and no treatment). Situations such as these give rise to the terms **false positive, false negative, true positive, and true negative**. Several examples are given as they apply to medical tests, lie detectors and drug tests. This is a very practical unit that contains some very subtle ideas.

Key Words and Phrases

false positive   false negative   true positive

true negative

Key Concepts and Skills

• Explain the issues that can occur when data are averaged.

• Compute percentages in tables and interpret them correctly.

• Understand the subtleties in interpreting medical, drug, and polygraph tests.

4   Managing Money

Overview

One of the most immediate ways in which mathematics affects every person's daily routine is through finances: bank accounts, credit cards, loans, investments, and income taxes. In this chapter we take an in-depth look at these personal financial matters. In addition, we will explore the financial affairs of the United States as we look at the federal budget.

Unit 4A Taking Control of Your Finances

Overview

The key to your financial success is to approach all of your financial decisions with a clear understanding of the available choices. This unit focuses on maintaining a personal **budget**, for without a clear picture of your budget and monthly **cash flow**, you cannot effectively plan for the future.

A four-step budget is outlined in the text. Following this simple procedure allows you to take control of your personal finances. The four steps are:
- List your monthly income, including prorated amounts for income that is not received on a monthly basis.
- List *all* your monthly expenses, including prorated amounts for occasional expenses.
- Compute your cash flow.
- Make adjustments as needed to improve your cash flow.

Monthly *cash flow* is computed by subtracting your monthly expenses from your monthly income. With a positive cash flow, you'll have money left over at the end of the month (which can be applied toward savings); with a sustained negative cash flow, you are headed toward financial ruin.

Key Words and Phrases

    budget   cash flow

Key Concepts and Skills

- Understand the importance of controlling your finances.
- Know how to make a budget and compute cash flow.
- Be aware of your bank balance, know what you spend, and don't overspend your budget.

Unit 4B The Power of Compounding

Overview

The phenomenon of **compounding** plays a fundamental role in all of finance, both on the investment side of the coin (earning and saving money) and on the loan side (borrowing money).

The unit opens by considering simple and compound interest problems as they arise in banking. Compound interest problems rely on four pieces of information:

- **starting principal**, which we call P,
- **annual percentage interest rate**, which we call APR,
- **number of compoundings per year**, which we call n, and
- **number of years** the account is held, which we call Y.

With these variables defined, the compound interest formula is written as

$$A = P\left(1 + \frac{\text{APR}}{n}\right)^{(nY)}.$$

This formula (and others in the chapter) must be evaluated on a calculator. Be sure you study the highlighted boxes that show you how to use your calculator effectively; needless to say, practice helps immensely.

An important distinction must be made between the APR and the **annual percentage yield** (APY) of a bank account. If an account compounds interest more than once per year, then the balance will increase by *more* than the APR in one year (due to the power of compounding). The percent by which the balance in the account actually increases in a year is the APY.

The more often compounding takes place during the year, the more the balance increases. The limiting case occurs when compounding takes place every instant, or **continuously**. With continuous compounding, you get the maximum return on your money (for a fixed APR). The continuous compounding formula, $A = P \times e^{(\text{APR} \times Y)}$, involves the mathematical constant e, which is approximately 2.71828. You should become familiar with how to compute various powers of e on your particular calculator.

The unit concludes with a very practical problem. The usual compound interest formulas tell you how much money you will have in your account after Y years with a given initial deposit *today*. But what if you know you would like to have, say $30,000, in 20 years? How much should you deposit today in order to reach this goal? This is an example of a **present value** problem, and it's quite important for planning purposes.

Key Words and Phrases

simple interest	principal, P	compound interest
annual percentage rate (APR)	compound interest formula	annual percentage yield (APY)
continuous compounding	present value, PV	

Key Concepts and Skills

- Determine the balance in an account with simple interest given an initial deposit and interest rate.
- Determine the balance in an account with compounding given an initial deposit, APR, number of compoundings, and number of years.

- Determine the balance in an account with continuous compounding given an initial deposit, APR, and number of years.
- Understand the difference between APR and APY, and be able to compute the APY for an account.
- Determine the present value for an account that will produce a given balance after a specified number of years.

Important Review Boxes

- Brief Review: Powers and Roots
- Brief Review: Four Basic Rules of Algebra

Unit 4C Savings Plans and Investments

Overview

An **annuity** is a special kind of savings plan in which deposits are made regularly, perhaps every month or every year. People who are creating a college fund or building a retirement plan usually use an annuity. An annuity account increases in value due to compounding (as in regular bank accounts studied in the previous unit) and due to the regular deposits. Not surprisingly, the mathematics of annuity plans follows naturally from the compound interest formulas of the previous unit.

There is just one basic formula that needs to be mastered in order to work with annuities:

$$A = \text{PMT} \times \frac{\left[\left(1 + \dfrac{\text{APR}}{n} \right)^{(nY)} - 1 \right]}{\left(\dfrac{\text{APR}}{n} \right)}.$$

This formula uses the variables needed for the compound interest formula of the previous unit (APR, number of compoundings, and number of years), *plus* the amount of the regular deposits (PMT). To keep matters simple (but still realistic) we assume that the regular deposits are made as often as interest is compounded. There is no doubt that the savings plan formula is complicated. Be sure to study the *Using Technology* box to learn how to manipulate your calculator when evaluating the formula.

Just as with a bank account, it is practical to ask present value questions with savings plans. If you know how much money you would like to have at a future time (perhaps the time of retirement), then how much should you deposit monthly between now and then in order to reach that goal? Several practical examples of present value problems are presented in the text.

Having discussed the mathematics of investments, the unit concludes with a comprehensive summary of various types of investment plans. Some key considerations in choosing investment plans are **liquidity** (how easily your money can be withdrawn), **risk** (how safe your money is), and **return** (the amount that you earn on your investment). Liquidity is fairly easy to determine, risk is less easy to assess, and for return, there are simple formulas that give the total return and annual return on an investment:

$$\text{total return} = \frac{(A-P)}{P}, \quad \text{and} \quad \text{annual return} = \left(\frac{A}{P}\right)^{(1/Y)} - 1.$$

We examine four general types of investments in light of these three factors:
- stocks,
- bonds, and
- Treasury bills.

These three types of investment have very different characteristics in terms of liquidity, risk, and return.

An essential part of investing money is keeping track of how well your investments are doing. Many people follow their investments using the financial pages of the newspapers. For this reason, we spend some time discussing how to read the financial pages for stocks, bonds, and mutual funds. Bonds in particular have some special terminology (such as **face value**, **coupon rate**, and **discount**) and a special formula for calculating the **current yield** on a bond.

Key Words and Phrases

savings plan	annuity	savings plan formula
stocks	bonds	cash
mutual fund	liquidity	risk
return	total return	annual return
portfolio	shares	market price
dividends	P/E ratio	percent yield
face value	coupon rate	maturity date
discount	points	load vs. no load
current yield		

Key Concepts and Skills

- Determine the value of a savings plan given the regular deposit, APR, number of compoundings, and number of years.
- Determine the present value for a savings plan that will yield a given balance after a specified number of years.
- Understand the terms liquidity, risk, and return as they apply to various forms of investment.
- Determine the annual yield on an investment given the total return.
- Understand the terms used to report stocks in the financial pages of a newspaper.
- Understand the terms used to report bonds in the financial pages of a newspaper.
- Understand the terms used to report mutual funds in the financial pages of a newspaper.

- Compute the current yield of a bond.

Unit 4D Loan Payments, Credit Cards, and Mortgages

Overview

While most people have bank accounts or investments that earn money, the reality is that most people also have debts due to loans of one kind or another. In this unit we will explore the most common kinds of loans: short-term loans (such as automobile loans), loans on credit cards, and **mortgages** (or house loans).

One basic formula governs all loan problems; it is called the **loan payment formula**:

$$\text{PMT} = \frac{P \times \left(\dfrac{\text{APR}}{n} \right)}{\left[1 - \left(1 + \dfrac{\text{APR}}{n} \right)^{(-nY)} \right]}.$$

This formula uses the now familiar variables of the previous units, though with different financial terms attached to some of them.
- PMT = regular payment amount,
- P = the initial amount of the loan (called the **loan principal**),
- APR = the interest rate on the loan,
- n = the number of payment periods per year, and
- Y = the loan term (the number of years over which it will be paid back).

Once again, the loan payment formula is rather complicated, so be sure to study the *Using Technology* box.

The entire unit is devoted to using the loan payment formula for various practical problems. There are a lot of strategic questions involved in choosing a loan. Often you must choose between a loan with a short term and a loan with a longer term (it is typical to find that longer term loans have slightly higher interest rates). The best choice is not always clear, though the examples provide a good starting point in the decision making process.

For most people, a mortgage is the largest loan entered into over the course of a lifetime. Since mortgages involve large amounts of money over long periods of time, the decisions you make in choosing a mortgage are critical. The options and strategies involved with mortgages are explored (specifically, the differences and advantages of fixed-rate mortgages over adjustable rate mortgages). Also discussed are issues of refinancing, points, and closing costs. All in all, this is a practical unit, whose lessons will serve you well.

Key Words and Phrases

loan principal	installment loan	loan payment formula
mortgage	down payment	fixed rate mortgage
adjustable rate mortgage	closing costs	points
prepayment penalties	refinance	

Key Concepts and Skills

- Know the terminology associated with loans.
- Determine the loan payment given the loan principal, the APR, the number of payment periods, and the term of the loan.
- Analyze two loan options to determine which is best for a given situation.
- Be familiar with the practical issues associated with a home mortgage, such as points, closing costs, and fixed rate vs. adjustable rate loans.

Unit 4E Income Taxes

Overview

Another aspect of most peoples' financial lives is income taxes. The mathematics of income taxes is not difficult. However there are a lot of concepts and terminology involved with the subject. This unit is designed to give you a fairly thorough survey of income taxes, enough for you to make sense of complex tax forms and make critical decisions.

The first step in determining the amount of income tax you owe is to calculate your **taxable income**. From your **gross income** (the total of all your income), you subtract deductions and exemptions to arrive at your taxable income. Having computed your taxable income, the next step is to determine the tax itself. The U.S. tax system uses tax brackets or **marginal tax rates**: the percentage of your taxable income that you pay depends on your tax bracket. It also depends on your **filing status**: whether you are filing as a single person, a married person, or a head of household. This is also an appropriate place to discuss the **marriage penalty**.

Two other forms of income tax are **Social Security taxes** and **Medicare taxes** (collectively called FICA taxes). These taxes apply to income from wages (not investments), and for self-employed people, these taxes must be paid in full by the taxpayer. The rules that govern these taxes are discussed with examples.

Of great political and personal interest are the taxes assessed on **capital gains**, profits made from selling property or stocks. We will examine the laws for capital gains and give several examples that illustrate the laws. Also of practical value are the tax implications of **tax-deferred savings plans** that many people use for retirement pensions and the **mortgage interest tax deduction**, which is a large tax benefit for people holding mortgages.

In this unit you will find some relief from the big formulas and long calculations of previous units in this chapter. This chapter is very conceptual and it presents a lot of new terminology and practical information.

Key Words and Phrases

gross income	adjusted gross income	deductions
exemptions	dependents	standard deduction
itemized deduction	taxable income	tax credits
progressive income tax	marginal tax rates	filing status
social security tax	Medicare tax	capital gain
ordinary income	short-term capital gain	long-term capital gain
tax-deferred savings plan	mortgage interest deduction	marriage penalty

Key Concepts and Skills

- Understand the basic terminology associated with income tax.
- Compute taxable income from gross income given deductions and exemptions.
- Use marginal tax rates to compute the tax on a given taxable income.
- Compute the social security and Medicare taxes on a given income.
- Understand and determine taxes on capital gains.
- Incorporate tax-deferred savings plans and mortgage interest deductions into a tax calculation.

Unit 4F Understanding the Federal Budget

Overview

Several times in this book, particularly in problems about large numbers, we have referred to the federal budget or the federal debt. There is no question that the federal budget is a source of truly huge numbers – numbers so large that most people, including politicians, don't even have a sense of their size.

In this chapter, we give a brief survey of the essentials of the federal budget. While the details can be confusing, the general principles of the budget are straightforward. The government has **receipts** or **income** (primarily from taxes) and it is has **outlays** or **expenses** (that cover everything from defense to education to social security). In a given year, if outlays exceed receipts, then the government has a **deficit**. On the other hand, if receipts exceed expenses, then the government has a **surplus**. If the government has a deficit then it must borrow money and in the process goes into **debt**. The devastating effect of going into debt is that **interest** must be paid on the borrowed money. This interest payment itself becomes an expense category in the budget. The U.S. government typically spends around 6% of its budget on interest on the debt. If the government has many consecutive deficit years (as has been the case in the United States), then the debt continues to grow and cannot be paid off until there are surplus years.

In this unit we clarify the ideas of deficit and debt by looking at the budget of an imaginary small business. We then consider the federal budget for the last several years and see how the major spending categories have contributed to a deficit in each year. With many deficit years, it also becomes clear how the debt has grown to its current level of about $17 trillion.

The problems with the Social Security system are explored, and these raise a subtle point in federal budget reports. If you look closely at the federal budget summaries in this unit, you will see that the increase in the debt from one year to the next is *not* equal to the deficit for the current year, as one might expect. The explanation lies in the distinction between the **publicly held debt** and **gross debt**, and the way in which the U.S. government divides its **unified budget** into two parts: the **off-budget** component (i.e. the Social Security system), and the **on-budget** component.

A balanced budget can be achieved either by increasing receipts or decreasing outlays or both. However, not all outlays can be reduced: **mandatory** expenses must be paid and **entitlements** are very difficult to cut. This leaves only **discretionary** expenses that are eligible for cuts.

The federal budget, together with its deficits, surpluses, and debts, affects everyone in many ways. This unit will give you a basic understanding of the important issues.

Key Words and Phrases

receipts	outlays	net income
deficit	surplus	debt
mandatory outlays	entitlements	discretionary outlays
publicly held debt	gross debt	unified budget
on budget	off budget	

Key Concepts and Skills

- Explain the terms receipts, outlays, surplus, deficit, and debt.
- Given a budget for a company or government, compute surplus or deficit.
- Explain the difference between publicly held debt and gross debt.
- Understand the major issues concerning the future of Social Security.

5   Statistical Reasoning

Overview

Much of the quantitative information that flows over us every day is in the form of data that are gathered through surveys and other statistical studies. When a person, business, or organization wants to know something, often the first idea that comes to mind is "collect data." From news reports to scientific research, from political polls to TV viewer surveys, we are surrounded by statistics. How are these numbers gathered? Are the conclusions based on those numbers reliable? And most important of all, should you believe the results of a statistical study? Should you change your life based on the results of a statistical study?

These are the questions that we will *begin* to answer in this chapter. The emphasis of this chapter is on qualitative issues, and there will not be a lot of computation involved. We will resume the study of statistical problems in Chapter 6 when we look at the quantitative side of statistics. If you master these two chapters, you will have a good foundation in statistical studies — enough to allow you to read the news critically and to take a complete course in statistics.

Unit 5A Fundamentals of Statistics

Overview

The goal of this unit is to learn how a statistical study should (and should *not*) be done. Of utmost importance is the distinction between the **population**, the group of people or objects that you would like to learn about, and the **sample**, the group of people and objects that you actually measure or survey. There are basically two aspects to any statistical study:
- collecting data from the people or objects in the sample, and
- drawing conclusions about the entire population based on the information gathered from the sample.

There are many ways that either of these steps can go wrong. Most of this chapter will deal with the first step, which is often called **sampling**. This step is crucial; if a sample is not chosen so that it is representative of the entire population, the conclusions of the study cannot be reliable.

Four different sampling methods are discussed:
- simple random sampling
- systematic sampling
- convenience sampling
- stratified sampling

It is also important to distinguish between two basic types of statistical studies. An **observational study** is one in which participants (people or objects) are questioned, observed, or measured, but in no way manipulated. By contrast, in an **experiment**, the participants *are* manipulated in some way, perhaps by dividing them into a **treatment group** and a **control group**. A subtle borderline situation is a **case-control study**, in which participants naturally form two or more groups by choice (for example, people who choose to smoke and those who choose to abstain). While a case-control study looks like an experiment, it is actually an observational study.

Other features of statistical studies that should be noted are the **placebo effect** (when people who *think* they received a treatment respond as if they received the treatment, even though they did not), **single blinding** (when participants do not know whether they receive a treatment), and **double blinding** (when neither researchers nor participants know who received a treatment).

You often hear statements like, "The President's approval rating is 54% with a margin of error of 5%." Because results of surveys are often given in this way, we will also look briefly at the notions of **confidence interval** and **margin of error**. Although these ideas are rather technical, it is important to understand how to interpret them. In Chapter 6, we will see how confidence intervals and margins or error are actually determined.

Key Words and Phrases

sample	population	raw data
sample statistics	population parameters	representative sample
simple random sampling	systematic sampling	convenience sampling
stratified sampling	bias	observational study
experiment	treatment group	control group
placebo	placebo effect	single blind experiment
double-blind experiment	case control study	cases
controls	margin of error	confidence interval

Key Concepts and Skills

- Understand the distinction between a sample and the population.
- Know the five basic steps of a statistical study.
- Be able to form a representative sample by various sampling methods.
- Understand the difference between an observational study, an experiment, and a case-control study.
- Understand the placebo effect, and single and double blinding.
- Interpret margins of error and confidence intervals.

Unit 5B Should You Believe a Statistical Study?

Overview

Suppose you read that 68% of all TV viewers watched the final game in the NCAA basketball tournament, or that 1 in 5 people in the world do not have access to fresh drinking water, or that a pre-election poll states that the Republican candidate leads by 5 percentage points. In this unit, we address a basic question: how do you know whether to believe the claims of these statistical studies?

The unit takes the form of a list of eight guidelines for evaluating a statistical study. Each guideline is accompanied by examples and analyses. It may not be possible to assess a particular study in light of *all* eight guidelines, but the more of these guidelines that a study satisfies, the more confident you can be about the conclusions of the study. The eight guidelines are listed below.

1. Get a Big Picture View: Identify the goal, type and population of the study.

2. Consider the source.

3. Look for bias in the sample.

4. Look for problems in defining or measuring the variables of interest.

5. Beware of confounding variables.

6. Consider the setting and wording in surveys.

7. Check that results are presented fairly.

8. Stand back and consider the conclusions.

Key Words and Phrases

bias   variables   selection bias

participation bias

Key Concepts and Skills

- Understand the eight guidelines for evaluating a statistical study, and be able to carry them out on a particular study.

Unit 5C Statistical Tables and Graphs

Overview

We often think of data and statistics as long lists of numbers, and indeed they start in that form. But these numbers don't become meaningful until they are summarized in some digestible form, and one of the best ways to represent long lists of numbers is with a picture. In this and the next unit, we will explore the many ways in which quantitative information can be displayed. This unit focuses on the basic types of graphs, ones that can often be drawn with a pencil and paper. The next unit highlights the more exotic types of graphs which are more difficult to produce, but equally common in news and research reports.

Often the best first step in summarizing data is to make a **frequency table** that shows how often each category in a study appears (for example, the number of students receiving grades of A, B, C, D, and F on an exam). Such a table can also show the **relative frequency** and the **cumulative frequency** for each category.

The type of display used for a set of data also depends on the type of data that are collected. We distinguish between **qualitative data** (non-numerical categories that have no natural order) and **quantitative data** (numerical values that are ordered).

Here is a summary of the types of graphs that will be considered in this unit.

- **Bar graphs** are used to show how some numerical quantity (for example, population) varies from one category to another (for example, for several different countries). The categories are usually qualitative and can be shown in any order.

- **Pie charts** are used to show the fraction (or percentage) of a population that falls into various categories. The categories are usually qualitative and can be shown in any order. For example, a pie chart could be used to show the fraction of a class that has brown, black, blond, or red hair. Recall the total pie must always represent the total relative frequency of 100%.

- **Histograms** are bar graphs in which the categories are quantitative, and thus have a natural order. Histograms often show how many objects or people are in various categories. For example, a histogram would be used to show the number of people in a town that fall into age categories 0–9, 10–19, 20–29, and so on.

- **Line charts** serve the same purpose as histograms, but instead of using bars to indicate the number of objects or people in each category, they use dots connected by lines.

- **Time-series diagrams** are usually histograms or line charts that show how a quantity changes in time. For example, the day-to-day changes in the stock market could be displayed as a time-series diagram.

Key Words and Phrases

frequency table	relative frequency	cumulative frequency
bar graph	pie chart	histogram
line chart	time-series diagram	

Key Concepts and Skills

- Construct a frequency table for a set of table showing frequencies, relative frequencies, and cumulative frequencies.
- Determine an appropriate kind of display for a given set of data.
- Display data with a bar graph by finding the correct height of the bars.
- Display data with a pie chart by finding the correct angles for the sectors of the pie.
- Display data with a histogram by finding the correct height of the bars.
- Display data with a line chart by finding the correct location of points on the line chart.
- Display data with a time-series diagram by finding the correct location of points on the diagram.

Unit 5D Graphics in the Media

Overview

Whereas the previous unit dealt with graphs that are relatively easy to produce yourself, this unit surveys more complicated displays that frequently appear in the media. The goal is not to actually produce these more sophisticated graphs; often a powerful software packages is needed to create them. Rather the emphasis will be on interpreting the information presented in a graphic.

Here is a summary of the types of graphs that we will consider in this unit.
- **Multiple bar graphs** are used to show how *two or more* quantities (for example, population and birth rate) vary from one category to another (for example, for several different countries). Multiple bar graphs are really two or more bar graphs combined with each other; they include multiple histograms. **Multiple line charts** may also be used.
- **Stack plots** are used to display several quantities simultaneously, often showing how they all change in time. For example, a stack plot could be used to show how the incidence of several diseases has changed in time.
- **Graphs of geographical data**, such contour plots and weather maps, give a two-dimensional picture of how a quantity varies over a geographical region. Such displays are also used to show the distribution of a disease over a geographical region.
- **Three-dimensional** graphs take many forms, but they are all used to show how three different quantities are related to each other.

As graphical displays become more complex, there is more opportunity for confusion and deception. The unit closes by discussing several ways in which you can be misled by such displays — they must be interpreted with caution.

Key Words and Phrases

multiple bar and line graphs	stack plots	three-dimensional graphics
contour plots	exponential scale	pictographs
perceptual distortions	percent change graphs	

Key Concepts and Skills

- Interpret multiple bar and line graphs and histograms.
- Interpret stack plots.
- Interpret three-dimensional graphics.
- Interpret contour plots.
- Interpret percent change graphs.
- Interpret exponential scales.
- Detect deceptive pictographs or other displays of data.

Unit 5E Correlation and Causality

Overview

This unit deals with the critical task of determining whether one event *causes* another event. The process often begins by looking for **correlations** between two variables; a correlation exists when higher values of one variable are consistently associated with higher (or lower) values of the second variable. For example, body weight is correlated with height, because in general, greater body weights go hand-in-hand with greater heights. Correlations can be detected by making a **scatter diagram** of the two variables in question.

There are three possible explanations for a correlation between two variables. It may be due to
- a coincidence,
- a common underlying cause, or
- a genuine cause and effect relation.

Of the three possibilities, the most interesting and the most difficult to establish is the last one: when can we be sure that one event caused another?

The unit presents six methods (attributed to the philosopher John Stuart Mill) for establishing cause and effect relations:

> 1. Look for situations in which the effect is correlated with the suspected cause even while other factors vary.

> 2. Among groups that differ only in the presence or absence of the suspected cause, check that the effect is similarly present or absent.

> 3. Look for evidence that larger amounts of the suspected cause produce larger amounts of the effect.

> 4. If the effect might be produced by other potential causes (besides the suspected cause), make sure that the effect still remains after accounting for these other potential causes.

> 5. If possible, test the suspected cause with an experiment. If the experiment cannot be performed with humans for ethical reasons, consider doing the experiment with animals, cell cultures, or computer models.

> 6. Try to determine the physical mechanism by which the suspected cause produces the effect.

Because causality is important in legal cases, we also look at the legal standards for establishing causality. They are crucial in determining the guilt or innocence of defendants. They are
- possible cause,
- probable cause, and
- cause beyond reasonable doubt.

The unit also presents two case studies in which identifying a cause and effect relation led to a discovery or the solving of a mystery.

Key Words and Phrases

correlation	scatter diagram	positive correlation
negative correlation	coincidence	underlying cause
causality	probable cause	possible cause
cause beyond reasonable doubt		

Key Concepts and Skills

- Construct a scatter diagram for two variables.
- Identify positive, negative, and no correlation between two variables.
- Understand the three explanations for a correlation.
- Apply the six methods for determining causality.
- Understand the three legal levels of confidence in causality.

6 Putting Statistics to Work

Overview

We now continue our study of statistics, having already introduced some of the qualitative aspects in Chapter 5. The treatment of statistics in this chapter is more quantitative and involves some numerical calculations. In Unit 6A we present several methods for characterizing data sets, including the calculation of the mean, median, and mode. The next unit continues in the same vein by discussing measures of variation in a data set. In Unit 6C, we introduce the normal distribution which lies at the heart of many statistical methods. The final unit discusses the powerful process known as *statistical inference* – drawing reliable conclusions about a population based on data taken from a sample of that population.

Unit 6A Characterizing Data

Overview

Typical data sets often contain hundreds or thousands of numbers, so one goal is to summarize data sets in compact and meaningful ways. In this unit we explore methods used to describe and summarize data beyond the frequency tables discussed in the previous chapter.

For a very concise summary of a set of data, it is common to compute its **mean**, **median**, and/or **mode**. These one-number summaries are defined as follows:

- The *mean* of a data set is calculated by the formula

$$\text{mean} = \frac{\text{sum of all values}}{\text{total number of values}}.$$

- The *median* is the middle score in the data set. Note that there will be two "middle" values if a data set has an even number of data points; if the two middle values are different, the median lies halfway between them.
- The *mode* is the most common score in a data set. A data set may have more than one mode, or no mode.

Having determined where the "center" of the data set lies, it is next useful to describe the shape of the data: are the data values symmetric? Are they weighted to the right or left? These questions can be answered qualitatively by finding whether the data values are **positively** or **negatively skewed**. How do outliers affect the mean and median?

Key Words and Phrases

distribution	mean	median
mode	single-peaked	bimodal
symmetric	outliers	left-skewed
right-skewed	variation	

Key Concepts and Skills

- Compute the mean, median, mode of a set of data.
- Understand the effect of outliers on mean and median.
- Be aware of possible sources of confusion about averages.
- Determine the qualitative shape of a data set and identify whether it is skewed.
- Understand the basic idea of what variation represents.

Unit 6B Measures of Variation

Overview

We continue the theme of the previous unit by exploring additional ways to summarize a distribution of data in compact ways. In this unit, we look at methods to describe the *spread* or **variation** of a set of data. The simplest measure of variation is called the **range**, the difference between the maximum and minimum data values. We can improve on the range by computing the **upper** and **lower quartiles**, the mean, and the maximum and minimum data values to make a **five-number summary** for the distribution. A **boxplot** is a picture of the five-number summary and provides a good snapshot of the distribution.

The variation is best captured in a single number by computing the **standard deviation** of a distribution, shown here:

$$\text{standard deviation} = \sqrt{\frac{\text{sum of (deviations from the mean)}^2}{\text{total number of data values} - 1}}.$$

This calculation can become laborious for large data sets, and is often done with calculators or computer software. But you should know how to set up the standard deviation calculation using a table, as shown in Table 6.2. A useful approximation to the standard deviation is given by the **range rule of thumb**: $\text{standard deviation} \approx \dfrac{\text{range}}{4}$.

Key Words and Phrases

range	five-number summary	upper quartile
lower quartile	box plot	deviation
standard deviation		

Key Concepts and Skills

- Compute the range of a data distribution.
- Find the lower and upper quartiles of a data distribution.
- Construct the five-number summary and a boxplot for a data distribution.
- Compute and interpret the standard deviation of a small data set.

Unit 6C The Normal Distribution

Overview

This unit is about the normal distribution, which plays an important role in all of statistics. This unit lays the groundwork for the following unit in which we learn how to draw conclusions about an entire population when we have data from a relatively small sample of that population. This process of inferring something about the population based on a sample is often called **inferential statistics**.

We begin by introducing the **normal distribution**, a whole family of symmetric distributions that are characterized by their mean and their standard deviation. Many variables, such as test scores, heights, weights, and other physical characteristics, follow a normal distribution closely. One of the most important results about the normal distribution is the **68-95-99.7 Rule:**

- About 68% (closer to 68.3%) of the data points fall within 1 standard deviation of the mean.
- About 95% (closer to 95.4%) of the data points fall within 2 standard deviations of the mean.
- About 99.7% of the data points fall within 3 standard deviations of the mean.

Thus if we know that a variable is distributed normally and we know its mean and standard deviation, we can say a lot about where a particular data value lies in the distribution. Working with the normal distribution is made much easier with the use of **standard scores** (or **z-scores**) and **percentiles**. The standard score simply measures how many standard deviations a particular data value is above or below the mean. Once the standard score is known, the percentile can be found using Table 6.3 (which appears in all statistics books). Similarly, given a percentile, the corresponding z-score can be found.

Key Words and Phrases

normal distribution	68-95-99.7 rule	percentile
standard (z-) score		

Key Concepts and Skills

- Identify normal distributions and know how they arise.
- Draw a normal distribution curve with a given mean and standard deviation.
- Use the 68-95-99.7 Rule to analyze a normal distribution.
- Convert standard scores to percentiles and vice versa.

Unit 6D Statistical Inference

This unit explains in a fairly qualitative way one of the most important aspects of statistics: the process by which information from a sample can be used to make reliable claims about the entire population from which the sample is taken. This is a technical topic, but it is so important that it's worth understanding, at least in general terms.

The unit begins with an explanation of **statistical significance**: a set of observations or measurements is said to be statistically significant if it is unlikely to have occurred by chance. For example, if we toss a fair coin 100 times and see 52 heads, we would not be surprised, and might attribute the outcome to chance. However, if we observed 80 heads in 100 tosses of a fair coin, it would be difficult to attribute this outcome to chance; we would say this outcome is statistically significant.

We can be more precise if we use probability in an intuitive way:
- if the probability of an observed outcome occurring by chance is 1 in 20 (0.05) or less, the outcome is significant at the **0.05 level of significance.**
- if the probability of an observed outcome occurring by chance is 1 in 100 (0.01) or less, the outcome is significant at the **0.01 level of significance**.

The discussion next returns to **margins of error** and **confidence intervals**. The problem of interest is very common for opinion polls and surveys. Suppose you want to know the proportion (or percentage) of people in a population who hold a particular opinion or have some trait. The usual practice is to select a sample and determine what proportion (or percentage) of people in the sample hold that particular opinion or have that trait; this is called the **sample statistic**. Knowing the sample statistic, what can you conclude about the entire population?

You will see in the unit (with plenty of explanation and examples) that if the sample has n individuals, then the proportion of interest for the population lies within $1/\sqrt{n}$ of the sample statistic — with 95% certainty. The quantity $1/\sqrt{n}$ is called the **margin of error** and the interval that extends $1/\sqrt{n}$ on either side of the sample statistic is called the **95% confidence interval**. For example, if in a randomly selected sample of $n = 10,000$ people, 45% have blond hair, then the margin of error is $1/\sqrt{10,000} = 0.01 = 1\%$ and the confidence interval is $45\% - 1\%$ to $45\% + 1\%$ or 44% to 46%. We can conclude with 95% certainty that between 44% and 46% of the entire population has blond hair. This is one of the most basic results of inferential statistics. It allows us to say something about an entire population based on what we learn from a sample.

The last topic in the unit is hypothesis testing, which is the procedure for determining whether a claim about a population, based on observations of a sample, is valid. The claims we will study involve either a population mean (for example, the mean weight of all college women is 125 pounds) or a population proportion (for example, 54% of all college students eat pizza at least once a week). The hypothesis test involves two hypotheses. The **null hypothesis** claims a specific value for a population parameter (it is often the value expected in the case of no special effect). It takes the form:

null hypothesis: population parameter = claimed value

The **alternative hypothesis** is the claim that is accepted if the null hypothesis is rejected.

There are two possible outcomes of a hypothesis test:
- rejecting the null hypothesis, in which case we have evidence that supports the alternative hypothesis.
- not rejecting the null hypothesis, in which case we lack sufficient evidence to support the alternative hypothesis.

How do we decide whether or not to reject the null hypothesis? Here's where the subject gets a bit technical to carry out in detail. But we can say it in words:

- If the chance of a sample result at least as extreme as the observed result is less than 1 in 100 (or 0.01), the test is significant at the 0.01 level. The test offers strong evidence for rejecting the null hypothesis (and accepting the alternative hypothesis).
- If the chance of a sample result at least as extreme as the observed result is less than 1 in 20 (or 0.05), the test is significant at the 0.05 level. The test offers moderate evidence for rejecting the null hypothesis.
- If the chance of a sample result at least as extreme as the observed result is greater than 1 in 20, the test is not significant. The test does not provide sufficient grounds for rejecting the null hypothesis.

While this subject is challenging, the text does offer several good examples and analogies to clarify this important topic.

Key Words and Phrases

statistical significance	sampling distribution	confidence interval
margin of error	null hypothesis	alternative hypothesis
significance at the 0.05 level	significance at the 0.01 level	

Key Concepts and Skills

- Explain the meaning of statistical significance.
- Construct and interpret a confidence interval given a margin or error.
- Describe the basic purpose and aspects of a hypothesis test.
- Construct a null and alternative hypothesis for a given situation.
- Given the relevant probability, decide whether or not to reject a null hypothesis.

7   Probability: Living With the Odds

Overview

Probability is involved in nearly every decision we make. Often it is used on a subjective or intuitive level; but occasionally we try to be more precise. As you will see in this chapter, probability is one of the most applicable branches of mathematics. Before we are done with the chapter, we will be able to apply probability to lotteries, gambling, life insurance problems, and air traffic safety. It is a fascinating chapter, full of mathematics and real-life problems.

Unit 7A Fundamentals of Probability

Overview

We all have an intuitive sense of what a probability is and that sense is useful. A probability is a number between zero and one. If the probability of an event is zero, then it cannot happen; if the probability of an event is one, then it is certain to happen; and if the probability is somewhere in between zero and one, it gives a measure of how likely the event is to happen.

We can approach probabilities in three different ways:

- A **theoretical probability** uses mathematical methods to find the probability of an event occurring.
- A **relative frequency probability** (or **empirical probability**) is determined by making many observations and counting the number of times an event occurs.
- A **subjective probability** is based on intuition.

This unit focuses on theoretical probability calculations, with a lighter treatment of empirical and subjective probabilities.

If you want to find the probability that a fair die will show a 5 when rolled once, you might reason that there are six possible **outcomes** of a single roll, and a success (rolling a 5) is one of those six outcomes. You (correctly) conclude that the probability of rolling a 5 is 1/6. This is the basic procedure for computing theoretical probabilities:

Step 1: Count the number of possible outcomes of an **event**.
Step 2: Count the number of outcomes that represent **success** — that is, the number of outcomes that represent the sought after result.
Step 3: Determine the probability of success by dividing the number of successes by the total number of possible outcomes:

$$P(A) = \frac{\text{number of outcomes that represent success}}{\text{total number of possible outcomes}}.$$

This procedure can be used to compute probabilities in many different situations where the outcomes are equally likely, as you will see in the examples of the text.

An often-used important fact is that if the probability of an event A occurring is $P(A)$, then the probability that it does *not* occur is $1 - P(A)$. For example, the probability of rolling a fair die and getting a five is 1/6; therefore the probability of rolling *anything but* a five is $1 - 1/6 = 5/6$.

Suppose you flip three coins and are interested in the number of heads that appear. In this case, there are four possible events: three heads, two heads, one head, and three tails. We will show how to find the probability of all of the events. The collection of the probabilities of *all* possible events is called a **probability distribution**.

There are many situations in which it is impossible to determine theoretical probabilities. In these situations, we can't use mathematical methods, but we have records or data to work with. Such an approach leads to empirical probabilities. For example, we often talk about a 100-year flood, which means that based on historical records, a flood of this magnitude occurs once every 100 years. So the empirical probability of such a flood occurring in any given year is 1/100. We discuss empirical probabilities because they are used so often in practice.

Finally, we take a few pages to clarify the confusion that can arise over the use of the word *odds*. The odds *for* an event A are found by dividing the probability of an event occurring by the probability of the event *not* occurring. For example, the odds of rolling a fair die and getting a five are *1 to 5*, because $(1/6) \div (5/6) = 1/5$. The odds *against* an event A are found by dividing the probability of an event not occurring by the probability that it does occur, which is just the reciprocal of the odds for an event. Thus the odds against rolling a five are *5 to 1*.

Key Words and Phrases

outcome	event	theoretical probability
relative frequency probability	subjective probability	probability distribution
probability of an event not occurring	odds	

Key Concepts and Skills

- Distinguish between theoretical probability, empirical probability, and subjective probability.
- Use the three-step process to determine simple theoretical probabilities.
- Find probability distributions for coin and dice experiments.
- Determine empirical probabilities from data or historical records.
- Find the probability of an event *not* occurring.
- Find the odds for an event from its probability.

Important Review Box

- Brief Review: The Multiplication Principle

Unit 7B Combining Probabilities

Overview

In the previous unit, we studied methods for finding the probability of individual events. However, many interesting situations actually consist of multiple events. For example, what is the probability of tossing ten heads in a row with a fair coin? Or what is the probability of drawing a jack or a heart from a standard deck of cards? We will answer these and many other practical questions in this unit.

The unit presents five different techniques that involve multiple events. As always, it's important to know *how* to apply these techniques and *when* to apply them. The five situations covered by these methods are:

- independent AND events
- dependent AND events
- either/or mutually exclusive events (non-overlapping)
- either/or jointly possible events (overlapping)
- *at least once* rule for independent events.

The term **AND event** refers to two or more events *all* happening. For example, what is the probability of rolling two fair dice and seeing a six on *both* dice (a six *and* a six)? Or what is the probability of drawing four cards from a standard deck and getting an ace each time? The important distinction with AND events is whether they are **independent** or **dependent**. If the occurrence of one event does not affect the others (for example, in rolling two dice, the outcome of one die does not affect the outcome of the other die), then we have independent events. The rule that applies for two independent events is

$$P(A \text{ and } B) = P(A) \times P(B).$$

This principle can be extended to any number of independent events. For example, the AND probability of A *and* B *and* C is

$$P(A \text{ and } B \text{ and } C) = P(A) \times P(B) \times P(C).$$

If the events are not independent, then a bit more care is needed. Suppose you draw two cards from a deck, but don't replace the first card before drawing the second. The outcome of the second card depends on the outcome of the first card. The rule that applies in this case for two dependent events is

$$P(A \text{ and } B) = P(A) \times P(B \text{ given } A).$$

With **either/or events**, we are interested in the probability of *either* event A *or* event B occurring. The important distinction with either/or events is whether the events in question are mutually exclusive. If the occurrence of the first event prevents the occurrence of the second event, then the events are **mutually exclusive** or **non-overlapping**. For example, when selecting one card from a standard deck, drawing a heart and drawing a diamond are mutually exclusive events because if one event occurs (say, drawing a heart), then the other event (drawing a diamond) cannot occur. The rule that applies in this case with two mutually exclusive events is

$$P(A \text{ or } B) = P(A) + P(B).$$

This principle can be extended to any number of mutually exclusive events. For example, the probability that either A *or* B *or* C occurs is

$$P(A \text{ or } B \text{ or } C) = P(A) + P(B) + P(C).$$

If events are **overlapping** (for example, going into a room of people and meeting a woman *or* a Democrat), then a modification of the above rule must be used. The rule that applies in this case is

$$P(A \text{ or } B) = P(A) + P(B) - P(A \text{ and } B).$$

In general, for AND probabilities we multiply probabilities and for either/or probabilities we add probabilities. However, modifications must be made in the cases of dependent or overlapping events.

Finally, a very important situation arises when we ask about the probability of an event happening *at least once* in a string of several independent trials of an experiment. For example, if you buy ten lottery tickets, what is the probability of *at least* one ticket being a winner? If you roll a die ten times, what is the probability of rolling *at least* one six? The rule that gives the probability of an event *A* occurring *at least once* in *n* trials is

$$P(A \text{ at least once in } n \text{ trials}) = 1 - P(A \text{ does not occur in } n \text{ trials})$$

$$= 1 - P(\text{not A})^n.$$

The unit has explanations of these rules and many examples to show how they are used.

Key Words and Phrases

AND probability	independent events	dependent events
either/or probability	overlapping events	non-overlapping events
at least once rule		

Key Concepts and Skills

- Identify AND probabilities and either/or probabilities.
- Distinguish between independent events and dependent events.
- Distinguish between non-overlapping events and overlapping events.
- Identify *at least once* situations.
- Know when and how to apply the five probability rules given in the unit.

Unit 7C The Law of Large Numbers

Overview

This unit is designed to strengthen your intuition about probabilities and to point out some common misconceptions about probabilities. It also deals with the concept of expected value, which has many important applications.

We start by discussing the **law of large numbers**. We know that the probability of tossing a head with a fair coin is 1/2. This does not mean that when you toss a coin twice, you will always get one head and one tail. Nor does it mean that when you toss a coin ten times, you will always get five heads and five tails. What we *can* say is that the more often you toss the coin, the closer the proportion of heads is to the probability of finding heads on a single coin toss. This is an example of the law of averages.

The important concept in this chapter is **expected value**. Consider a situation in which there are several different outcomes. Each outcome has a known probability and a known payoff, or **value** (which could be positive or negative). For example, a lottery may have a $10 prize, a $100 prize, a $1000 prize, and a grand prize of $1,000,000. These are the values (payoffs) of the four outcomes, and each outcome has a certain probability. If you play this lottery many times, how much do you expect to win or lose in the long run? The answer is given by the expected value.

There is a general rule for computing expected values. In a situation with just two outcomes, the expected value is given by

$$\text{expected value} = \binom{\text{value of}}{\text{event 1}} \times \binom{\text{probability of}}{\text{event 1}} + \binom{\text{value of}}{\text{event 2}} \times \binom{\text{probability of}}{\text{event 2}}.$$

The unit presents several other applications of expected value as it arises in life insurance policies, lotteries, and the **house edge** in casino gambling. There are some important practical lessons to be learned here.

The law of averages is related to the principle called the **gambler's fallacy**. Gamblers often feel that if they get behind or start losing money, then their luck will change and their chances of winning will improve. The fact is that probabilities do not change just because someone is losing. Once behind, it is likely that you will stay behind or get even further behind.

Key Words and Phrases

law of large numbers   expected value   gambler's fallacy

house edge

Key Concepts and Skills

- Explain the law of averages.
- Explain the gambler's fallacy.
- Compute the expected value in situations with two or more outcomes.
- Understand the house edge and compute it in specific situations.

Unit 7D Assessing Risk

Overview

This unit is a nice collection of topics, all related in some way to probability. The first topic is estimating risk. You often hear statements such as "you are more likely to get struck by lightning than die in an airplane crash." How are such statements concocted? And to what extent are they true? We will explore how such comparisons of risk are made, particularly as they apply to **vital statistics** — data about births and deaths.

You will see that these problems are examples of empirical probability. If we know the number of deaths due to, say, heart disease, we can estimate the probability of a person dying from heart disease. If we know the number of deaths due to automobile accidents, we can estimate the chances of a person dying in an automobile accident. With these two facts, it is possible to compare the risks of automobiles and heart disease. Clearly, risk analysis and decision-making are closely related.

Related to vital statistics are the subjects of **mortality** and **life expectancy**. We will look at graphs of death rates and life expectancy as they change with age. The interpretation of these graphs reveals a few surprises.

Key Words and Phrases

accident rate   death rate   vital statistics

life expectancy   mortality

Key Concepts and Skills

- Determine the probability of an event given the number of people who experience the event.
- Understand different measures of risk.
- Compare the risks of two events or two causes of death.
- Interpret life expectancy and mortality tables.

Unit 7E Counting and Probability

Overview

We all know how to count *individual* objects or people, but there are many other types of counting problems that arise when we want to count *groups* of objects or people. That is the subject of this unit. There are four basic counting methods presented in this chapter:

- selections from two or more groups (multiplication principle),
- arrangements with repetition,
- permutations (where repetition is not allowed), and
- combinations (again, repetition is not allowed).

In this unit, we will learn not only how to apply these methods, but equally important, how to determine *when* to use each method.

Selection from two or more groups occurs when we have several different groups of objects and we need to select one item from each group. For example, in a restaurant, you might have four choices for an appetizer (first group), six choices for a main course (second group), and five choices for a dessert (third choice). How many different meals could you select?

Problems of this sort can be solved using a table, a tree, or (the easiest of all) the **multiplication principle** (see Review Box in Unit 7A). The total number arrangements of items from two or more groups is

(number of items in first group) × (number of items in second group) × (number of items in third group) × ···· × (number of items in last group).

In the above example, there would 4 × 6 × 5 = 120 different possible meals that you could order.

Arrangements with repetition occur when you select items from a single group and items may be used more than once. Often the items in problems of this kind are letters or numerals. For example, how many different three-digit area codes can be formed from the numerals 0 through 9? There are 10 ways to choose the first numeral and, because repetition is allowed, there are 10 ways to choose the second numeral, and 10 ways to choose the third numeral. This amounts to a total of 10 × 10 × 10 = 1000 different three-digit area codes. The general rule is that

if we make r selections (for example, the three digits of the area code) from a group of n items (for example, the numerals $0 - 9$), n^r different arrangements are possible.

Permutations also involve selecting items from a single group with one important difference — repetition is not allowed. Furthermore, in counting permutations, the order of the arrangements matters; that is, we count ABCD and DCBA as two different arrangements (or permutations). To summarize, permutations require

- selection from a single group,
- repetition is not allowed, and
- order matters.

In the unit, we work through several examples towards the general formula for counting permutations. We need a bit of mathematical notation along the way, so we introduce the **factorial** function.

$$n! = n \times (n-1) \times (n-2) \times \ldots \times 2 \times 1.$$

With factorials in hand, we can write a general formula for counting permutations.

If we make r selections from a group of n items, the number of possible permutations is

$$_nP_r = \frac{n!}{(n-r)!},$$

where $_nP_r$ is read as "the number of permutations of n items taken r at a time."

Combinations are much like permutations with one difference — order of selection does not matter; that is, we count ABCD and DCBA as the *same* combination. The requirements for combinations are

- selection from a single group,
- repetition is not allowed, and
- order does *not* matter.

The number of combinations of *n* objects taken *r* at a time is the same as the number of permutations, except that we have to correct for the overcounting embedded in the permutation formula (permutations involve a specific order and combinations don't). This subtle point is explained in the text. The result is the combinations formula.

If we make *r* selections from a group of *n* items, the number of possible *combinations*, in which order does not matter, is

$$ _nC_r = \frac{_nP_r}{r!} = \frac{n!}{(n-r)! \times r!} $$

where $_nC_r$ is read as "the number of combinations of *n* items taken *r* at a time."

The key to mastering these methods is knowing not only *how* to use a particular method, but *when* to use it. There are plenty of practice exercises at the end of the unit; you should work as many as possible. Another helpful hint is to be certain you are using your calculator effectively. Most calculators compute factorials directly, and some have keys for permutations and combinations — it is in your best interest to become familiar with them.

The unit concludes with a brief look at the subject of **coincidence**. Should you be surprised during an evening of playing cards when you are dealt a 13-card hand with ten hearts? Many so-called coincidences are bound to happen to someone, but we are often surprised when they happen to us. The famous birthday problem at the end of the unit illustrates this principal. We also explore the phenomenon of **streaks**, particularly in sports, when a player repeats a certain success for many consecutive games.

Key Words and Phrases

selection from two or more groups	multiplication principle	arrangements with repetition
permutation	combination	coincidence

Key Concepts and Skills

- Describe the four different counting methods discussed in the unit.
- Determine which of the four methods applies in a counting problem.
- Apply each of the four methods on appropriate problems.
- Understand why not all coincidences should be surprising.

Important Review Box

- Brief Review: Factorials

---

8   Exponential Astonishment

Overview

If you had to learn just one lesson from a quantitative reasoning course, it might well be the difference between exponential growth (the present chapter), and linear growth (studied in Chapter 9). Exponential growth and decay impact our everyday lives in many ways, but most people are not aware of its presence or its power. In this chapter, you will learn how bank accounts, populations, and radioactive waste are governed by exponential models.

Unit 8A Growth: Linear versus Exponential

Overview

The difference between linear and exponential growth is stated at the outset.

- *Linear growth* occurs when a quantity grows by the same *absolute* amount in each unit of time.
- *Exponential growth* occurs when a quantity grows by the same *relative* amount — that is, by the same *percentage* — in each unit of time.

Notice how these facts are related to the ideas of absolute and relative change studied in Unit 3A. This unit introduces the ideas surrounding exponential growth in a very intuitive fashion. We use three parables (*From Hero to Headless*, *The Magic Penny*, and *Bacteria in a Bottle*) that illustrate quite dramatically the power of exponential growth. The goal is to develop some intuition about exponential growth and understand doubling processes.

These lessons are summarized in the highlight box at the end of the unit:
- Exponential growth is characterized by repeated doublings. With each doubling the amount of increase is approximately equal to the *sum* of all preceding doublings.
- Exponential growth cannot continue indefinitely. After only a relatively small number of doublings, exponentially growing quantities reach impossible proportions.

Key Words and Phrases

   linear growth   exponential growth   doublings

Key Concepts and Skills

- Explain the difference between linear and exponential growth.
- Identify whether a given growth pattern is linear or exponential.
- Understand the implications of a doubling process.

Unit 8B Doubling Time and Half-Life

Overview

Exponential growth is characterized by a constant **doubling time**. If a quantity (for example, a population or a bank account) is growing exponentially, then it doubles its size during a fixed period of time, and it continues to double its size over that same period of time for as long as current growth rates continue. For example, if a tumor, growing exponentially, has a doubling time of two months, then it doubles its number of cells during the first two months, and doubles that number again during the next two months, and continues to double *every* two months. Knowing the doubling time essentially defines the growth pattern for all times.

We denote the doubling time with T_{double}. If we know T_{double} for a particular quantity that grows exponentially, then over a period of time t, the quantity will increase by a factor of

$$2^{t/T_{double}}.$$

If we know the doubling time and the initial value of a particular quantity that grows exponentially, then we can find its value at all later times. The new values at later times are given by

$$\text{new value} = \text{intial value} \times 2^{t/T_{double}}.$$

If we know that a quantity grows with a constant percentage growth rate (for example, 5% per year), then we know it grows exponentially and that it has a constant doubling time. This suggests that there should be a connection between the percentage growth rate P, and the doubling time. We use a specific example of an exponentially growing population to present a widely used formula that relates percentage growth rate to doubling time. It is called the **Approximate Doubling Time Formula** or the **Rule of 70**. It says that

$$T_{double} \approx \frac{70}{P}.$$

This formula is an approximation and works best when the percentage growth rate is small (say, less that 10%). For example, if a bank account grows at 4% per year, its balance will double in approximately 70/4 = 17.5 years.

Everything we learned about exponential growth has a parallel with exponential decay. For example, if a quantity decays exponentially at a rate of 5% per month, it *decreases* by 5% every month. A quantity that decays exponentially has a constant **half-life** — the period of time over which it decreases its size by 50% or one-half.

If we know T_{half} for a particular quantity that decays exponentially, then over a period of time t, the quantity will decrease by a factor of

$$\left(\frac{1}{2}\right)^{1/T_{half}}$$

If we know the half-life and the initial value of a particular quantity that decays exponentially, then we can find its value at all later times. The new values at later times are given by

$$\text{new value} = \text{inital value} \times \left(\frac{1}{2}\right)^{1/T_{half}}.$$

The **Approximate Half-Life Formula** also applies to exponentially decaying quantities. If a quantity decreases by $P\%$ per unit time, its half-life is given by

$$T_{half} = \frac{70}{P}.$$

As with the Approximate Doubling Time Formula, this half-life formula is approximate and works best when P is small (say, less than 10%).

For those who are curious, the unit closes with the *exact* doubling time and half-life formulas. These formulas are a bit more complicated and require the use of logarithms. They are exact for all percentage growth and decay rates, not just small ones.

Key Words and Phrases

doubling time	percentage growth rate	approximate doubling time formula
Rule of 70	half-life	percentage decay rate
approximate half-life formula	exact doubling time formula	exact half-life formula

Key Concepts and Skills

- Identify the percentage growth rate from the description of an exponential growth process.
- Find the doubling time from percentage growth rate.
- Find the percentage growth rate from doubling time.
- Determine the value of an exponentially growing quantity given the doubling time and initial value.
- Identify the percentage decay rate from the description of an exponential decay process.
- Find the half-life from percentage decay rate.
- Find the percentage decay rate from half-life.
- Determine the value of an exponentially decaying quantity given the half-life and initial value.

Important Review Box

- Brief Review: Logarithms

Unit 8C Real Population Growth

Overview

In the previous units of this chapter, we have seen that exponential growth models allow only for rapid and continual growth. While this sort of growth is realistic for some populations (for example, bacteria, tumor cells, or small populations of animals), it cannot continue forever. Eventually lack of space or resources must limit growth.

In this short unit we look briefly at more realistic approaches to population modeling. We begin by discussing how the overall growth rate really consists of **birth rates** and **death rates** (and immigration rates). We then introduce the important concept of **carrying capacity** — the maximum number of individuals that the environment can sustain. Any realistic population model must account for the carrying capacity.

The next observation is that as a population grows, its percentage growth rate cannot remain constant, as it does in an exponential growth model. The **logistic growth model** uses a growth rate that actually decreases as the population grows, and reaches zero when the population reaches the carrying capacity. An alternative scenario, called **overshoot and collapse**, is also presented as a realistic population model. The unit closes with a discussion of estimating the carrying capacity of the Earth.

Key Words and Phrases

birth rate	death rate	carrying capacity
logistic growth model	overshoot	collapse

Key Concepts and Skills

- Determine the exponential growth rate from birth and death rates.
- Understand the limitations of exponential growth models.
- Understand the assumptions and effects of a logistic growth model.
- Determine growth rates for a logistic model.
- Describe overshoot and collapse models.

Unit 8D Logarithmic Scales: Earthquakes, Sounds, and Acids

Overview

Closely related to exponential growth and decay models are **logarithmic scales** — these are just the other side of the exponential coin, in that they scale quantities based on powers (exponents).

Each of the three logarithmic scales explored in the text (the *earthquake magnitude scale*; the *decibel scale* for sound; and the *pH scale* for measuring acidity) takes two equivalent forms, one expressed in terms of logarithms, and one expressed in terms of powers. Following are the defining equations for these scales.

The earthquake magnitude scale:

$$\log_{10} E = 4.4 + 1.5M \quad \text{or} \quad E = (2.5 \times 10^4) \times 10^{1.5M} .$$

The decibel scale for sound:

$$dB = 10\log_{10}\left(\frac{\text{intensity of the sound}}{\text{intensity of the softest audible sound}}\right)$$

or

$$\frac{\text{intensity of the sound}}{\text{intensity of the softest audible sound}} = 10^{(\text{loudness in dB/10})}.$$

The pH scale:

$$pH = -\log_{10}[H^+] \quad \text{or} \quad [H^+] = 10^{-pH}.$$

One key idea to understand about logarithmic scales is that changing a scaled number by a multiplicative factor does *not* correspond to a change in the quantity the scale measures by the same multiplicative factor. For example, if the magnitude of an earthquake is doubled from 4 to 8, the energy E released is not doubled, but rather it goes up by a *much* larger factor (in this case, the energy released is about 1,000,000 times as great). The factor by which the measured quantity increases with corresponding increases in the scaled number depends upon how the logarithmic scale is defined.

An interesting aspect of the intensity of sound (measured by the decibel scale) is that it follows an **inverse square law**. This is just one of several quantities in nature that obey the inverse square law (others include gravity, and the brightness of light).

Key Words and Phrases

logarithmic scale	earthquake magnitude scale	decibel scale
pH scale	acid	base
neutral	acid rain	inverse square law

Key Concepts and Skills

- Determine the energy released by an earthquake given its magnitude, and compare two earthquakes of different magnitudes.
- Compare the intensity of sounds that have different decibel levels.
- Understand how the inverse square law affects the intensity of sound at a distance.
- Understand the relationship between pH and [H$^+$].
- Explore the relationship between the pH scale and acid rain.

9   Modeling Our World

Overview

Just as a map or a globe is a model of the earth and a set of floor plans is a model of a building, we can also create mathematical models that represent real-world problems and situations. In fact, it might be argued that one of the fundamental goals of mathematics is to create mathematical models.

In this chapter, we introduce perhaps the most basic tool of mathematical modeling: the function. In fact, we have already encountered modeling and functions prior to this chapter. The financial formulas of Chapter 4 are examples of functions, and the exponential formulas of Chapter 8 are examples of mathematical models. The presentation of functions in this chapter may be different than those you have seen elsewhere. We start at the beginning and develop the idea of a function in a practical and visual way. Hopefully, in this way, whether you are seeing functions for the first time or not, you will become comfortable with this essential mathematical concept.

Unit 9A Functions: The Building Blocks of Mathematical Models

Overview

This unit is brief and has a single purpose. Using words, a few definitions, pictures, and tables, the goal is to give a very qualitative idea of a **function**.

A function is a relationship between two quantities or **variables**. A function could describe how your height increases in time or how the temperature decreases with altitude. We call the variables in question the **independent variable** and the **dependent variable**. The terminology arises because we usually think of the dependent variable *changing with respect to* the independent variable; that is, if we make a change in the independent variable, it produces a change in the dependent variable.

There are three different ways to visualize or represent a function:
- with a data table of values of the two variables,
- with a graph (picture), and
- with an equation or formula.

In the remainder of this unit, we explore the first two ways of representing functions. The first approach (a data table) is straightforward and probably familiar. So we focus on graphing functions. The use of equations will be studied in the next unit.

Two concepts are important in graphing functions.

- The **domain** of a function is the set of values that both make sense and are of interest for the *independent variable*.
- The **range** of a function consists of the values of the *dependent variable* that correspond to the values in the domain.

If you can identify the domain and range of a function, then you have saved yourself a lot of work. You need only to graph those values of the independent variable in the domain and those values of the dependent variable in the range. Having defined the domain and range, we present several examples of functions and their graphs.

Key Words and Phrases

mathematical model	function	variable
independent variable	dependent variable	graph
coordinate plane	axis	origin
quadrants	domain	range

Key Concepts and Skills

- Find the domain and range of a function given in table form.
- Use the domain and range of a function to scale the axes of its graph.
- Know when it makes sense to "fill in" between the points of a graph.
- Use the four-step process to create a function's graph given a data table.

Important Review Box

- Brief Review: The Coordinate Plane

Unit 9B Linear Modeling

Overview

From the previous unit, you probably appreciate that graphs of functions can take many different forms. In this unit we study one very special, but widely used, family of functions — those functions whose graphs are straight lines. Not surprisingly, these functions are called **linear functions**.

The most important property of a linear function is its **rate of change**. If the graph of two variables is a straight line, it means that a fixed change in one variable always produces the *same* change in the other variable. The rate at which the dependent variable changes with respect to the independent variable is called the rate of change of the function. For linear functions, the rate of change is constant and it is equal to the **slope** of the graph. This fundamental property for linear functions is summarized here, in two equivalent ways.

$$\text{slope of a linear graph} = \frac{\text{vertical } \textit{rise}}{\text{horizontal } \textit{run}}$$

$$\text{rate of change} = \text{slope} = \frac{\text{change in } \textit{dependent variable}}{\text{change in } \textit{independent variable}}.$$

Having established the equivalence of the slope and the rate of change, we introduce a rule for calculating the change in the dependent variable. This rule simply says that if we know the rate of change of a linear function and we are given a change in the independent variable, then we can determine the corresponding change in the dependent variable:

Change in dependent variable = (rate of change) × (change in independent variable).

The rate of change rule is really just a stepping-stone to the final goal of the unit — to write a general equation for a linear function. After a detailed example to motivate the idea of a linear equation, we present the general form of a linear equation. It looks like this:

dependent variable = initial value + (rate of change × *independent variable*)

Up to this point we use words for the variable names. This practice can become cumbersome. So for the sake of economy, we can use single letter names for variables. For example, instead of writing *time*, we just use *t*; and instead of writing *number of chips*, we could use *N*. Don't let the use of letters confuse you; it just makes working with linear equations easier.

After several examples of creating and using linear equations, we come to one last topic. A linear equation is a model or a compact description of a particular situation. Once we have it, it can be used for prediction or to answer other useful questions. We want to be able to answer questions such as:

- When the independent variable has a certain value, what is the corresponding value of the dependent variable?
- When the dependent variable has a certain value, what is the corresponding value of the independent variable?

This brings us to the necessity of *solving* linear equations when we are given a particular value of either the dependent variable or the independent variable. Two simple rules are all we need to solve any linear equation for either variable:

- We can always add or subtract the same quantity from both sides of an equation.
- We can always multiply or divide both sides of an equation by a (nonzero) quantity.

Several examples conclude this action-packed unit. There is a lot of material in this unit and it is best to work at it slowly and allow plenty of time for reading and for practice.

Key Words and Phrases

linear function	rate of change	slope
initial value (*y*-intercept)	$y = mx + b$	

Key Concepts and Skills

- Determine the slope of a straight line given two points on the line.
- Determine the rate of change of a linear function, either from a description of the function or from two points associated with the function.
- Compute the change in the dependent variable given the rate of change and a change in the independent variable.
- Find the equation of a linear function given the rate of change and the initial value.
- Evaluate a linear function for the dependent variable given a value of the independent variable.
- Create a linear function from information about its rate of change and initial value.

- Solve a linear function for the independent variable given a value of the dependent variable.

Unit 9C Exponential Modeling

Overview

Having learned about the fundamentals of exponential growth and decay in Chapter 8, we can put these ideas to work to create models. Just as we used linear functions to model real world situations, we now use exponential functions to model situations in which exponential growth or decay occur. We begin by introducing the general **exponential function**

$$Q = Q_0 \times (1 + r)^t,$$

where t represents time. This equation requires an initial value Q_0 and a growth or decay rate r. Notice that if r is positive, then Q grows in time and if r is negative, then Q decays in time. Also important is that the units of time used for r and t must be the same (for example, if r is 5% *per day*, then t is measured in days). Once a specific exponential function is found, it can be used to predict the value of Q at all future times.

Be sure to study the highlight box entitled *Forms of the Exponential Function*. It shows that there are really several forms for the exponential function depending on whether you are given a growth rate, a doubling time, or a half-life.

With these exponential functions at hand, the rest of the unit is devoted to various applications. We look at how population growth, economic inflation, oil consumption, pollution, and drugs in the blood can all be modeled using these laws. Of particular importance is the technique called **radiometric dating**, which also relies on the exponential decay law. If these examples don't convince you of the widespread presence of exponential growth and decay, you will find even more applications in the problems.

Key Words and Phrases

exponential function   growth or decay rate   radiometric dating

Key Concepts and Skills

- Given either a growth rate or a doubling time, use the appropriate form of the exponential function to model an exponentially growing quantity.
- Given either a decay rate or a half-life, use the appropriate form of the exponential function to model an exponentially decaying quantity.
- Be familiar with forms of the exponential function that use the doubling time or half-life.
- Understand radiometric dating and know how to determine the age of a material that contains a radioactive element.

Important Review Box

- Brief Review: Algebra with Logarithms

10   Modeling with Geometry

If you like geometry, you will probably like this chapter. If you didn't, this chapter may show you some things you had not thought about before and may be able to use. The first unit is a review of some fundamental concepts from high school geometry.

The last two sections have applications many of you have not seen before. While the ideas in Unit 10B are not difficult, they are particularly useful in many walks of life. Applications to astronomy, road construction, and how tall to make a house under certain building restrictions are typical of the way simple geometry can be used to solve problems.

Unit 10C explores a few basic ideas underlying the relatively new and exciting area of mathematics called fractal geometry. You may have heard about fractals and even seen one (the front cover of your text is decorated with a fractal), but now you will know how to make your own, among other things.

Unit 10A Fundamentals of Geometry

Overview

Because much of this chapter is based on ideas and concepts from geometry, it makes sense to do some review. If much of this unit is familiar, just sit back and enjoy it.

The unit begins by defining and giving examples of the basic concepts of geometry: **points**, **lines**, **planes**, and **angles**. We then move on to familiar two-dimensional objects (also called plane objects). The most basic objects are **circles** and **regular polygons** such as **triangles**, **squares**, **pentagons**, and **hexagons**. You should know or be able to find quickly formulas for the
- area of a circle,
- circumference of a circle,
- area of a triangle, and
- area of a **parallelogram** (which include squares and rectangles).

We give several practical examples of the uses of these formulas.

The next subject is solid objects in three dimensions. You should be familiar with the table that has formulas for the
- surface area and volume of a **rectangular prism** (box),
- surface area and volume of a **cylinder** (soda can),
- surface area and volume of a **sphere**.

We give several examples of these formulas applied to practical problems.

The unit closes with a very useful and far-reaching section on scaling laws. We first discuss scale models, such as maps or architectural models. If we take an object and increase all of its dimensions by a factor of, say 10, then the **scale factor** is 10. The important result of this section is that
- areas scale with the *square* of the scale factor, and

- volumes scale with the *cube* the scale factor.

Thus, if you were to enlarge yourself by a scale factor of two, your surface area would increase by a factor of $2^2 = 4$, and your volume (and weight) would increase by a factor of $2^3 = 8$.

These scaling laws explain a lot of interesting phenomena. By considering the **surface-area-to-volume-ratio**, we can answer questions such as, why does crushed ice keep your drink colder than large ice cubes? Why can flies walk on ceilings? Why does the Moon have no active volcanoes?

Key Words and Phrases

geometry	Euclidean geometry	point
line	plane	dimension
angle	vertex	right angle
straight angle	acute angle	obtuse angle
perpendicular	parallel	radius
diameter	polygon	regular polygon
circumference	perimeter	parallelogram
rectangular prism	cube	cylinder
sphere	scale factor	scaling laws
surface to volume ratio		

Key Concepts and Skills

- Define and give examples of point, line, and plane.
- Convert angle measurements to fractions of a circle and vice versa.
- Determine the perimeter and area of common plane objects (circle, triangle, square, rectangle, parallelogram).
- Determine the surface area and volume of common three-dimensional objects (cube, rectangular prism, cylinder, sphere).
- Use geometrical ideas and formulas to solve practical problems.
- Understand scaling laws and surface-area-to-volume ratios.

Unit 10B Problem Solving with Geometry

Overview

Much of this chapter is based on ideas and concepts from geometry, but it involves applications which may be new to you. The unit begins by defining and giving examples of the basic concept of angles and how to measure them. The first application uses the perhaps familiar notion of **latitude** and **longitude**. As you may recall, this way of measuring the earth gives coordinates for locations with respect to the **prime meridian** and the **equator**. Knowing latitude and longitude, one can easily calculate the distance between two locations on the same meridian.

We explore the relationship between **angular size, physical size** and **distance** to an object. Using this relationship, one can calculate any one of the three from the other two. For example, one can estimate the diameter of the moon (its physical size) knowing the distance to it and its angular size.

The next application involves the notions of **pitch**, **grade**, and **slope.** They are basically the same thing, but one is usually preferred in a given context.

No discussion of geometry and measurement is complete without mentioning the **Pythagorean Theorem** and **similar triangles**. The first gives the familiar relationship between the lengths of the sides in a right triangle: $a^2 + b^2 = c^2$. The second gives a relationship between the sides of two triangles whose angles are all equal. This relationship is $\dfrac{a'}{a} = \dfrac{b'}{b} = \dfrac{c'}{c}$ where a, b, and c are the lengths of the sides in one triangle corresponding to the lengths a', b', and c' of the similar triangle. A nice application of similar right triangles is the calculation of heights from ground measurements such as the solar access problem of example 8.

The last section has to do with **optimal shapes**. As it turns out

- squares are optimal rectangles in that they maximize area while minimizing perimeter.
- cubes are optimal prisms in that they maximize volume while minimizing surface area.

Key Words and Phrases

latitude	longitude	prime meridian
equator	angular size	physical size
pitch	grade	slope
Pythagorean Theorem	similar triangles	optimal shape

Key Concepts and Skills

- Use latitude to find distance between two locations with same longitude.
- Use any two of angular size, physical size, and distance to an object to find the third.
- Find the slope from pitch or grade and/or find a distance or height from them.
- Find a distance using Pythagorean Theorem; also calculate the area of a right triangle.
- Identify similar triangles and use them to solve measurement problems.

Unit 10C Fractal Geometry

Overview

In this unit we explore perhaps the most recent mathematical development in this book. **Fractal geometry** arises with the observation that classical geometry (invented by the ancient Greeks and studied by all of us in high school) works very well for regular objects such as circles and squares. But many objects in the real world, particularly natural forms, are far more complex and are not described well by classical geometry. Fractal geometry was proposed, in part, to describe the complicated forms we find in nature.

We begin by looking at what happens when we measure the length of a line segment. If we continually magnify the line we don't see anything new. This means that if we measure the length of a line segment, we get the same length regardless of the length of the ruler. Similarly, if we measure the area of a square, we get the same area regardless of how small the ruler might be. And if we measure the volume of a cube, we get the same volume regardless of how small the ruler is. This leads to the conclusion that a line is a one-dimensional object, a square is a two-dimensional object, and a cube is a three-dimensional object.

Here is how we define the **dimension** of an object. We imagine successively reducing the length of our ruler by a **reduction factor** R. Each time we reduce the length of the ruler, we observe by what factor N the number of **elements** increases. In the case of regular objects, we find that…

- for a one-dimensional object (e.g., a line segment), $N = R^1$.
- for a two-dimensional object (e.g., a square), $N = R^2$.
- for a three-dimensional object (e.g., a cube), $N = R^3$.

For irregular objects, the result may be quite different. If we carry out the same measurement process, we may find that the relationship between R and N has the form $N = R^D$, but D is no longer a whole number. If D is not a whole number, the object is called a **fractal**.

With this new definition for the dimension of a fractal object in hand, we look at some examples of fractals, notably the **snowflake curve,** the **Cantor set,** the **Sierpinski triangle**, and the **Sierpinski sponge**. We also discuss how to measure the fractal dimension of realistic objects such as coastlines. We close with the most famous fractal of all, the **Mandelbrot set**.

Key Words and Phrases

fractal geometry	element	reduction factor
fractal dimension	snowflake curve	snowflake island
self-similar	Sierpinski triangle	Sierpinski sponge
Cantor set	Mandelbrot set	

Key Concepts and Skills

- Understand the meaning of dimension for regular objects (line, square, and cube).
- Understand the meaning of dimension for fractal objects.
- Determine the dimension of simple fractal objects.

11   Mathematics and the Arts

Overview

In this chapter we explore some very different applications of mathematics, namely music and the fine arts. As you will see, the connections between mathematics and the arts go back to antiquity. We devote one unit to music, one unit to classical painting, and the last to proportion as it appears in art and nature. This chapter should provide you with a new perspective and change of pace.

Unit 11A Mathematics and Music

Overview

The ties between mathematics and music go back to the ancient Greeks (~500 B.C.). The followers of Pythagoras discovered some of the basic laws that underlie our understanding of music today. They realized that the **pitch** of a musical note created by a plucked string is determined by the length of the string. This led to the more modern notions of the string's **fundamental frequency** (which measures how many times a string vibrates each second), and the principle that when the length of a string is halved, its pitch doubles (and goes up by an **octave**). With these few facts we can explain a lot.

In a standard scale that you might play on a piano or a guitar, one octave consists of 12 tones or **half-steps** (for example, the white and black keys between middle C and the next higher C). Here is the basic question we address: If we know the frequency of the first note of the scale, can we find the frequency of all 12 notes of the scale?

We show that to move up the scale a half step, we must *multiply* the frequency of the current tone by a fixed number. We give a brief argument showing that the magic number that generates the entire scale by multiplication is $f \approx 1.05946$, or the twelfth root of 2. It turns out the notes of the scale follow an exponential growth law, as discussed in Chapter 9.

When two notes, played together, have a pleasing sound (**consonant tones**), the ratio of their frequencies is found to be (nearly) the ratio of two small numbers, such as 3/2 or 4/3. We investigate how these ratios compare to the exact frequencies generated by the magic number f, and discuss the problem of *temperament*.

The unit closes with observations about the modern connections between music and mathematics: digital music, compact disks, and synthesizers.

Key Words and Phrases

sound wave	pitch	frequency
cycles per second	fundamental frequency	octave
scale	half-step	consonant tones
analog	digital	

Key Concepts and Skills

- Understand the relation between frequency and pitch.
- Determine the frequency of notes separated by an octave.
- Determine the frequency of notes on a 12-tone scale, given the frequency of the first note.
- Understand the difference between analog and digital music.

Unit 11B Perspective and Symmetry

Overview

It wasn't until the Renaissance (14th and 15th century) that painters attempted to draw three-dimensional objects realistically on a flat two-dimensional canvas. It took many years for these painters to perfect this technique, but in the end, they made a science of **perspective** drawing. In this unit we trace the development of perspective drawing and explore some of its mathematical necessities.

The key concept in perspective drawing is the **principal vanishing point**. It can be summarized as follows: If you are an artist standing behind a canvas, painting a real scene, then all lines that are parallel in the real scene and perpendicular to the canvas, must meet in a single point in the painting — this is the principal vanishing point. All other sets of parallel lines in the real scene (that are *not* perpendicular to the canvas) meet in their own vanishing points. All of the vanishing points (principal and otherwise) lie along a single line called the **horizon line**.

Another fundamental property of paintings and other objects of art is **symmetry**. Symmetry can mean many different things, but it often refers to a sense of balance. We define three different kinds of symmetry:

- **reflection symmetry**: an object can be "flipped" across a particular line and it remains unchanged (for example, the letter H).
- **rotation symmetry**: an object can be rotated through a particular angle and it remains unchanged (for example, the letter O).
- **translation symmetry**: an object or a pattern can be shifted, say to the right or the left, and it remains unchanged (for example, the pattern ....XXXXXX.... extended in both directions).

We investigate these symmetries in both geometrical objects and in actual paintings.

Symmetry arises in beautiful ways in **tilings** — patterns in which one or a few simple objects are used in repetition to fill a region of the plane. We give several examples of how triangles and quadrilaterals can be translated and reflected to produce wonderful patterns. You'll get the opportunity to try some tilings in the exercises.

Key Words and Phrases

vanishing point	principal vanishing point	horizon line
symmetry	reflection symmetry	rotation symmetry
translation symmetry	tiling	

Key Concepts and Skills

- Understand the role of the principal vanishing point in perspective drawing.
- Draw simple objects in perspective using vanishing points.
- Identify symmetries in simple objects.
- Draw simple objects with given symmetries.
- Create tilings from triangles or quadrilaterals using translations and reflections.

Unit 11C Proportion and the Golden Ratio

Overview

In this unit, we explore another fundamental aspect of art: proportion. Issues of proportion arise not only in the art created by humans, but in natural forms as well. The subject has a lot to do with aesthetics, our innate sense of what is beautiful. One of earliest statements about proportion and beauty goes back to the ancient Greeks, who defined and studied the **golden ratio**.

The first instance of the golden ratio arises in dividing a line segment. What division of a line segment has the most visual appeal and balance? The Greeks argued that the most visually appealing division is the one where

$$\frac{\text{length of long piece}}{\text{length of short piece}} = \frac{\text{length of entire piece}}{\text{length of long piece}},$$

that is, the ratio of the length of the long piece to the length of the short piece is the same as the ratio of the length of the entire segment to the length of the long piece. We can also write this as

$$\frac{L}{1} = \frac{L+1}{L}.$$

The value of L that solves this relationship is the special (irrational) number

$$\phi = \frac{1+\sqrt{5}}{2} = 1.61803\ldots$$

This number is called the **golden ratio** or **golden section**.

From the golden ratio, we can define a **golden rectangle**. Any rectangle with sides in the ratio of ϕ is called a golden rectangle. Both the golden ratio and the golden rectangle arise throughout the history of art. Great architectural works (for example, the Greek Parthenon) have dimensions close to those of the golden rectangle (although some claim that these examples are coincidences). Many common objects such as post cards and cereal boxes have dimensions of golden rectangles. We also look at some examples of how the golden rectangle appears in nature. There is one final connection that is too intriguing to ignore. We introduce the **Fibonacci sequence**, which was first used as a population model in the 13th century. Each term of the sequence is formed by adding the two previous terms:

$$1, 1, 2, 3, 5, 8, 13, 21, 34, \dots .$$

We show how this sequence is related, perhaps unexpectedly, to the golden ratio. And the circle is closed by observing that the Fibonacci sequence also appears in the artwork of nature. This is a unit with several seemingly unrelated ideas that eventually become linked in a beautiful way.

Key Words and Phrases

proportion	golden ratio	golden rectangle
logarithmic spiral	symmetry	Fibonacci sequence

Key Concepts and Skills

- Understand the golden ratio as a proportion and divide a line segment according to the golden ratio.
- Construct and identify golden rectangles.
- Generate the Fibonacci sequence and find the ratio of successive terms.
- Understand the connection between the golden ratio and the Fibonacci sequence.

12   Mathematics and Politics

Overview

This chapter explores the surprisingly crucial role that mathematics plays in political matters. The first two units deal with elections. In elections with more than two candidates, there are several ways to determine a winner, and they may not agree. In fact, it can be shown that there is no single voting method that meets certain fairness criteria. Unit 12C is devoted to another political process, namely apportionment. How do we determine the number of representatives that each state sends to Congress? Again, there are several methods that can be used, and they do not always agree. Finally, having discussed the problems of apportionment, the last unit discusses the contentious issue of redistricting.

Unit 12A Voting: Does the Majority Always Rule?

Overview

In this and the following unit we investigate mathematical problems associated with voting. This may sound like an unusual application of mathematics, but voting problems have been studied for several centuries and it has long been known that curious things can happen in voting systems.

The discussion begins with elections between two candidates. Throughout these units a candidate can be interpreted as a choice between two or more alternatives. For example, a candidate may be a person running for office or a brand of bagels in a taste test. With only two candidates, the outcome is straightforward: the **majority** rules. This means that the candidate with the most votes (which must be more than 50% of the vote) wins the election.

However, even with majority rule, there are some interesting situations that can arise. We first look at presidential elections in which the winner is chosen, not by the **popular vote**, but by the **electoral vote**. Historically, there have been U.S. presidential elections in which a candidate won the popular vote, but lost the election.

We then look at variations on majority rule that involve **super majorities**. For example, many votes in the U.S. government require more than a 50% majority: it takes a 2/3 super majority in both houses of Congress, followed by a 3/4 super majority vote of the states to amend the U.S. Constitution. More than a 50% majority of a jury is required to reach a verdict in a criminal trial (a unanimous decision is often required).

Things get interesting when we turn to elections with three or more candidates. Often such elections are based on a **preference schedule** in which each voter ranks the candidates in order of preference. For example, the following preference schedule shows the outcome of an election among five candidates that we call A, B, C, D, and E.

First	A	B	C	D	E	E
Second	D	E	B	C	B	C
Third	E	D	E	E	D	D
Fourth	C	C	D	B	C	B
Fifth	B	A	A	A	A	A
	18	12	10	9	4	2

There were a total of 55 voters and 18 voters ranked A first, D second, E third, C fourth, and B last. The other columns are interpreted in a similar way. The question is: how do we determine a winner of the election?

The remainder of the unit presents five methods to determine a winner to an election with a preference schedule.

- **plurality**: the candidate with the most first-place votes wins (candidate A would win the above election).
- **top-two runoff**: the top two candidates have a runoff in which the votes of the losing candidates are redistributed to the top two candidates (candidate B would win the above election).
- **sequential runoff**: the candidates with the fewest first place votes are successively eliminated one at a time, votes are redistributed, and runoff elections are held at each stage (candidate C would win the above election).
- **point system** (or **Borda count**): with five candidates, five points are awarded for each first place vote, four points are awarded for each second place vote, and so on. The candidate with the most points wins (candidate D would win the above election).
- **pairwise comparison**: the winner between each pair of candidates is determined and the candidate with the most pairwise wins is the winner of the election (candidate E would win the above election).

You can probably already see the dilemma. We have proposed five reasonable methods for finding a winner and they all give different results for the above preference schedule. The unit closes inconclusively with this question unanswered. The next unit takes up the issue of fairness in voting systems and attempts to resolve the question.

Key Words and Phrases

majority rule	popular vote	electoral vote
super majority	preference schedule	plurality
single runoff	sequential runoff	point system
Borda count	pairwise comparisons	filibuster

Key Concepts and Skills

- Understand the concept of majority rule and apply it in two-candidate elections.

- Know the difference between popular vote and electoral vote.
- Use super majority rules to determine the outcome of votes.
- Create and interpret preference schedules.
- Apply the methods of plurality, single runoff, sequential runoff, point system, and pairwise comparisons to determine the outcome of elections with preference schedules.

Unit 12B Theory of Voting

Overview

The previous unit closed with the observation that five reasonable methods for determining the winner of an election with three or more candidates can lead to five different winners. This unit attempts to resolve the dilemma, but as you will see, the resolution may be less than satisfying.

The analysis of voting methods involves what are known as **fairness criteria**. We will work with the following four criteria.

- **Criterion 1 (majority criterion):** If a candidate receives a majority of the first-place votes, that candidate should be the winner.
- **Criterion 2 (Condorcet criterion):** If a candidate is favored over every other candidate in pairwise races, then that candidate should be declared a winner.
- **Criterion 3 (monotonicity criterion):** Suppose that Candidate X is declared the winner of an election, and then a second election is held. If some voters rank X higher in the second election (without changing the order of other candidates), then X should also win the second election.
- **Criterion 4 (independence of irrelevant alternatives criterion):** Suppose that Candidate X is declared the winner of an election, and then a second election is held. If voters do not change their preferences, but one or more of the losing candidates drops out, then X should also win the second election.

These criteria have been developed by researchers who study voting methods; they are reasonable conditions that we would expect of any fair voting system. The first criterion, stating that a candidate with a majority of the votes should win, is the most natural. The second criterion is also straightforward: if a candidate has more votes than each of the other candidates taken individually, then that candidate should win the election. The third criterion just says that if some voters were to change their votes in favor of the winner of the election, then that candidate should still win the election. The last criterion says that if a losing candidate drops out of the election, the original winner should remain the winner.

Most of the unit is devoted to examining the five voting methods introduced in the previous unit to see whether they satisfy the four fairness criteria. You may be surprised at the results: none of the five methods always satisfies all of the fairness criteria. This is the message of a famous result in voting theory called **Arrow's Impossibility Theorem**. It says that there is no voting method that satisfies the four fairness criteria for all preference schedules.

We next look at a method called **approval voting** that has been proposed as an alternative way to handle elections with many candidates. It departs from the familiar one-person-one-vote idea that underlies all of the voting systems discussed so far (and it, too, has its drawbacks).

The unit closes with some ideas about power blocks in voting. An example of this idea is the formation of **coalitions,** which is common not only in our own government, but even more so in foreign governments.

Key Words and Phrases

fairness criteria	Arrow's Impossibility Theorem	approval voting
voting power	coalition	

Key Concepts and Skills

- Understand the four fairness criteria.
- Apply the four fairness criteria to each of the five voting methods (introduced in Unit 12A) for a given preference schedule.
- Understand the benefits and disadvantages of approval voting.
- Analyze an election in terms of coalitions.

Unit 12C Apportionment: The House of Representatives and Beyond

Overview

The U.S. Constitution stipulates that each state should have representation in the House of Representatives proportional to its population. But it doesn't specify exactly how the number of representatives should be determined. The process of assigning representatives according to population, which has applications beyond the House of Representatives, is called **apportionment**.

Apportionment begins by computing a **standard divisor**, which is the average number of people in the population per representative. For example, if the total population is 100,000 and there are 100 representatives, then the standard divisor is 1000 people per representative. The next step is to compute the **standard quota** for each state, which is the number of representatives a state should have if fractional representatives were possible. For example, if a state has 5500 people, then with a standard divisor of 1000, that state should have 5500/1000 = 5.5 representatives. Now the dilemma of apportionment can be seen: it is not possible to have fractions of representatives.

The founding fathers proposed several methods to overcome this problem of fractional representatives. Here is a quick survey of the various methods that have been used over the years (and those considered in detail in the text).

- **Hamilton's method**: First, round the standard quota for each state *down* to form the **minimum quota** for that state; then give any remaining representatives to the states with the largest fraction of representatives.

- **Jefferson's method**: Choose a **modified divisor** such that when the new standard quotas are rounded *down* (to form a **modified quota**), all the representatives are used.
- **Webster's method**: Choose a **modified divisor** such that when the new standard quotas are rounded *according to the standard rounding rule* (to form a **modified quota**), all the representatives are used.
- **Hill-Huntington method**: Choose a **modified divisor** such that when the new standard quotas are rounded *according to a modified rounding rule* (to form a **modified quota**), all the representatives are used.

Notice that the methods appear quite similar, but involve slightly different rules for rounding. However, as the examples in the text show, the methods can produce different results. Many examples are provided to show how these methods are carried out.

Which method is best? Two hundred years of American history have proved that none of the methods is perfect. Each method can exhibit at least one of several paradoxes or can violate certain reasonable expectations. To address this question of fairness, the unit explores each of the following situations:

- the **Alabama paradox**: when the total number of representatives increases, but at least one state actually loses representatives.
- the **population paradox**: when the population increases and under reapportionment a slow-growing state gains representatives at the expense of a fast-growing state.
- the **new states paradox**: when additional representatives are added to accommodate a new state and the apportionment of existing states changes.
- the **quota criterion**: under any apportionment, the number of representatives for any state should be one of the whole numbers nearest its standard quota (the standard quota rounded up or rounded down).

The apportionment problem is a bit like the voting problem, in that a perfect apportionment method cannot be found. This is the conclusion of the **Balinsky and Young theorem**.

Key Words and Phrases

apportionment	standard divisor	standard quota
Hamilton's method	Alabama paradox	population paradox
new states paradox	Jefferson's method	modified divisor
modified quota	quota criterion	Webster's method
Hill-Huntington method	geometric mean	Balinsky and Young theorem

Key Concepts and Skills

- Understand the problem and the dilemma of apportionment.
- Apply the four methods of the text to apportionment problems.
- Identify the deficiencies of the four methods in terms of paradoxes and the quota criterion.

Unit 12D Dividing the Political Pie

Overview

The difficulties of apportionment, discussed in Unit 12C, do not stop after it is determined how many House seats each state is to acquire. Within each state, a political battle is waged every ten years (after each census) to decide how to divide the state into its congressional districts. This process, called **redistricting**, is most easily described as a mapping problem, where each state must redraw previous district lines to reflect shifts in the population, the goal being to divide the state into *contiguous* regions (reflecting the number of representatives) of nearly equal population.

In most states, redistricting decisions are made in the legislature, and this is where politics and mathematics meet. When a particular party has a majority in the state legislature, it often uses its political clout to draw lines that result in favorable outcomes for that party. Only a loose set of rules, set by recent U.S. Supreme court decisions, must be followed when redistricting, and thus politicians go to great lengths to create districts that will win their party as many House seats as possible. The practice of drawing district boundaries for political advantage is called **gerrymandering**. (See Figure 12.4 for an example.)

The text provides several simplified examples of how political wrangling of this sort can very much influence both the number of seats a particular party wins, and the competitiveness of an election in a given district. Indeed, a striking graphic in Table 12.22 clearly shows a reduction in competitive seats in the House over the last four decades.

The unit closes with ideas for reforming the redistricting system. Ultimately, the solution may lie in the hands of mathematicians, who could be called upon to provide a fair algorithm for redistricting.

Key Words and Phrases

    redistricting   gerrymandering   contiguous

Key Concepts and Skills

- Understand the political and mathematical nature of redistricting.
- Be aware of how redistricting affects the competitiveness of an election.
- Understand the principals behind gerrymandering.
- Apply the legal requirements for redistricting: equal populations in each district, and contiguous districts.

UNIT 1A

QUICK QUIZ

1. **a.** By the definition used in this book, an argument always contains at least one premise and a conclusion.

2. **c.** By definition, a fallacy is a deceptive argument.

3. **b.** An argument must contain a conclusion.

4. **a.** Circular reasoning is an argument where the premise and the conclusion say essentially the same thing.

5. **b.** Using the fact that a statement is unproved to imply that it is false is appeal to ignorance.

6. **b.** "I don't support the President's tax plan" is the conclusion because the premise "I don't trust his motives" supports that conclusion.

7. **b.** This is a personal attack because the premise (I don't trust his motives) attacks the character of the President, and says nothing about the substance of his tax plan.

8. **c.** This is limited choice because the argument does not allow for the possibility that you are a fan of, say, boxing.

9. **b.** Just because A preceded B does not necessarily imply that A caused B.

10. **a.** By definition, a straw man is an argument that distorts (or misrepresents) the real issue.

DOES IT MAKE SENSE?

5. Does not make sense. Raising one's voice has nothing to do with logical arguments.

7. Makes sense. A logical person would not put much faith in an argument that uses premises he believes to be false to support a conclusion.

9. Does not make sense. One can disagree with the conclusion of a well-stated argument regardless of whether it is fallacious.

BASIC SKILLS AND CONCEPTS

11. a. *Premise:* Apple's iPhone outsells all other smart phones. *Conclusion:* It must be the best smart phone on the market.

    b. The fact that many people buy the iPhone does not necessarily mean it is the best smart phone.

13. a. *Premise:* Decades of searching have not revealed life on other planets. *Conclusion:* Life in the universe must be confined to Earth.

    b. Failure to find life does not imply that life does not exist.

15. a. *Premise*: He refused to testify. *Conclusion*: He must be guilty.

    b. There are many reasons that someone might have for refusing to testify (being guilty is only one of them), and thus this is the fallacy of limited choice.

17. a. *Premise:* Senator Smith is supported by companies that sell genetically modified crop seeds. *Conclusion:* Senator Smith's bill is a sham.

    b. A claim about Senator Smith's personal behavior is used to criticize his bill.

19. a. *Premise:* Good grades are needed to get into college, and a college diploma is necessary for a good career. *Conclusion:* Attendance should count in high school grades.

    b. The premise (which is often true) directs attention away from the conclusion.

21. False   23.   False

FURTHER APPLICATIONS

25. *Premise:* Obesity among Americans has increased steadily, as has the sale of video games. *Conclusion:* Video games are compromising the health of Americans. This argument suffers from the **false cause** fallacy. It's true obesity and video game sales have increased steadily for the last decade, but we cannot conclude that the latter caused the former simply because they happened together.

27. *Premise:* All the mayors of my home town have been men. *Conclusion:* Men are better qualified for high office than women. The conclusion has been reached with a **hasty generalization**, because a small number of male mayors were used as evidence to support a claim about all men and women.

29. *Premise:* My baby was vaccinated and later developed autism. *Conclusion:* I believe that vaccines cause autism. **False cause** is at play here, as the vaccination may have nothing to do with the development of autism, even though both are occurring at the same time.

31. *Premise:* Shakespeare's plays have been read for many centuries. *Conclusion:* Everyone loves Shakespeare. Both the premise and conclusion say essentially the same thing; this is **circular reasoning**.

33. *Premise:* After I last gave to a charity, an audit showed that most of the money was used to pay its administrators in the front office. *Conclusion:* I will not give money to the earthquake relief effort. This is a **personal attack** on charities. It can also be seen as an **appeal to ignorance**: the lack of examples of charities passing donations on to the intended recipients does not mean that a charity will not pass on donations.

35. *Premise:* The Congressperson is a member of the National Rifle Association. *Conclusion:* I'm sure she will not support a ban on assault rifles. This is a **personal attack** on members of the National Rifle Association. The argument also distorts the position of the National Rifle Association (not all members would oppose a ban on assault rifles); this is a **straw man**.

37. *Premise*: The Republicans favor repealing the estate tax, which falls most heavily on the wealthy. *Conclusion*: Republicans think the wealthy aren't rich enough. (Implied here is that you should vote for Democrats). The argument distorts the position of the Republicans; this is a **straw man**.

39. *Premise*: My boy loves dolls, and my girl loves trucks. *Conclusion*: There's no truth to the claim that boys prefer mechanical toys while girls prefer maternal toys. Using one child of each gender to come up with a conclusion about all children is **hasty generalization**. It can also be seen as an **appeal to ignorance**: the lack of examples of boys enjoying mechanical toys (and girls maternal toys) does not mean that they don't enjoy these toys.

UNIT 1B

QUICK QUIZ

1. **c**. This is a proposition because it is a complete sentence making a claim, which could be true or false.

2. **a**. The truth value of a proposition's negation (*not p*) can always be determined by the truth value of the proposition.

3. **c**. Conditional statements are, by definition, in the form of *if p, then q*.

4. **c**. The table will require eight rows because there are two possible truth values for each of the propositions *x*, *y*, and *z*.

5. **c**. Because it is not stated otherwise, we are dealing with the inclusive *or* (and thus either *p* is true, or *q* is true, or both are true).

6. **a**. The conjunction *p and q* is true only when both are true, and since *p* is false, *p and q* must also be false.

7. **b**. This is the correct rephrasing of the original conjunction.

8. **c**. This is the *contrapositive* of the original conjunction.

9. **b**. Statements are logically equivalent only when they have the same truth values.

10. **a**. Rewriting the statement in *if p, then q* form gives, "if you want to win, then you've got to play."

DOES IT MAKE SENSE?

7. Does not make sense. Propositions are never questions.

9. Makes sense. If restated in *if p, then q* form, this statement would read, "If we catch him, then he will be dead or alive." Clearly this is true, as it covers all the possibilities. (One could argue semantics, and say that a dead person is not caught, but rather discovered. Splitting hairs like this might lead one to claim the statement does not make sense).

11. Does not make sense. Not all statements fall under the purview of logical analysis.

BASIC SKILLS AND CONCEPTS

13. Since it's a complete sentence that makes a claim (whether true or false is immaterial), it's a proposition.

15. No claim is made with this statement, so it's not a proposition.

17. Questions are never propositions.

19. Asia is not in the northern hemisphere. The statement is false; the negation is true.

21. The Beatles were not a German band. The statement is false; the negation is true.

23. Sarah did go to dinner.

25. The Congressman voted in favor of discrimination.

27. Paul appears to support building the new dorm.

29. This is the truth table for the conjunction *q and r*.

q	r	q and r
T	T	T
T	F	F
F	T	F
F	F	F

31. "Cucumbers are vegetables" is true. "Apples are fruit" is true. Since both propositions are true, the conjunction is true.

33. "The Mississippi River flows through Louisiana" is true. "The Colorado River flows through Arizona" is true. Since both propositions are true, the conjunction is true.

35. "Some people are happy" is true (in general), as is "Some people are short," so the conjunction is true.

37. This is the truth table for *q and r and s*.

q	r	s	q and r and s
T	T	T	T
T	T	F	F
T	F	T	F
T	F	F	F
F	T	T	F
F	T	F	F
F	F	T	F
F	F	F	F

39. *Or* is used in the exclusive sense because you probably can't wear both a skirt and a dress.

41. The exclusive *or* is used here as it is unlikely that the statement means you might travel to both countries during the same trip.

43. *Or* is used in the inclusive sense because you probably would be thrilled to attend both concerts or the theater while in New York.

45. This is the truth table for the disjunction *r or s*.

r	s	r or s
T	T	T
T	F	T
F	T	T
F	F	F

47. This is the truth table for *p and (not p)*.

p	not p	p and (not p)
T	F	F
F	T	F

49. This is the truth table for *p or q or r*.

p	q	r	p or q or r
T	T	T	T
T	T	F	T
T	F	T	T
T	F	F	T
F	T	T	T
F	T	F	T
F	F	T	T
F	F	F	F

51. "Oranges are vegetables" is false. "Oranges are fruits" is true. The disjunction is true because a disjunction is true when at least one of its propositions is true.

53. "The Nile River is in Africa" is true. "China is in Europe" is false. The disjunction is true because a disjunction is true when at least one of its propositions is true.

55. "Trees walk" is false. "Rocks run" is also false. Since both are false, the disjunction is false.

57. This is the truth table for *if p, then r*.

p	r	if p, then r
T	T	T
T	F	F
F	T	T
F	F	T

59. *Hypothesis:* Eagles can fly. *Conclusion:* Eagles are birds. Since both are true, the implication is true, because implications are always true except in the case where the hypothesis is true and the conclusion is false.

61. *Hypothesis:* London is in England. *Conclusion:* Chicago is in Bolivia. Since the hypothesis is true, and the conclusion is false, the implication is false (this is the only instance when a simple *if p, then q* statement is false).

63. *Hypothesis:* Pigs can fly. *Conclusion:* Fish can brush their teeth. Since the hypothesis is false, the implication is true, no matter the truth value of the conclusion (which, in this case, is false).

65. *Hypothesis:* Butterflies can fly. *Conclusion:* Butterflies are birds. Since the hypothesis is true, and the conclusion is false, the implication is false (this is the only instance when a simple *if p, then q* statement is false).

67. If it rains (p), then I get wet (q).

69. If you are eating (p), then you are alive (q).

71. If you are bald (p), then you are a male (q).

73. *Converse:* If José owns a Mac, then he owns a computer. *Inverse:* If José does not own a computer, then he does not own a Mac. *Contrapositive:* If José does not own a Mac, then he does not own a computer. The converse and inverse are always logically equivalent, and the contrapositive is always logically equivalent to the original statement.

75. *Converse:* If Teresa works in Massachusetts, then she works in Boston. *Inverse:* If Teresa does not work in Boston, then she does not work in Massachusetts. *Contrapositive:* If Teresa does not work in Massachusetts, then she does not work in Boston. The converse and inverse are always logically equivalent, and the contrapositive is always logically equivalent to the original statement.

77. *Converse:* If it is warm outside, then the sun is shining. *Inverse:* If the sun is not shining, then it is not warm outside. *Contrapositive:* If it is not warm outside, then the sun is not shining. The converse and inverse are always logically equivalent, and the contrapositive is always logically equivalent to the original statement.

FURTHER APPLICATIONS

79. If you die young, then you are good.

81. If a free society cannot help the many who are poor, then it cannot save the few who are rich.

83. "If Sue lives in Cleveland, then she lives in Ohio," where it is assumed that Sue lives in Cincinnati. (Answers will vary.) Because Sue lives in Cincinnati, the hypothesis is false, while the conclusion is true, and this means the implication is true. The converse, "If Sue lives in Ohio, then she lives in Cleveland," is false, because the hypothesis is true, but the conclusion is false.

85. "If Ramon lives in Albuquerque, then he lives in New Mexico" where it is assumed that Ramon lives in Albuquerque. (Answers will vary.) The implication is true, because the hypothesis is true and the conclusion is true. The contrapositive, "If Ramon does not live in New Mexico, then he does not live in Albuquerque", is logically equivalent to the original conditional, so it is also true.

87. "If it is a fruit, then it is an apple." (Answers will vary.) The implication is false because, when the hypothesis is true, the conclusion may be false (it could be an orange). In the converse, "If it is an apple, then it is a fruit.", when the hypothesis is true, the conclusion is true, and this means the implication is true.

89. Believing is sufficient for achieving. Achieving is necessary for believing.

91. Forgetting that we are One Nation Under God is sufficient for being a nation gone under. Being a nation gone under is a necessary result of forgetting that we are One Nation Under God.

93. Following is a truth table for both *not* (p and q) and (*not* p) *or* (*not* q).

p	q	p and q	*not* (p and q)	(*not* p) *or* (*not* q)
T	T	T	F	F
T	F	F	T	T
F	T	F	T	T
F	F	F	T	T

Since both statements have the same truth values (compare the last two columns of the table), they are logically equivalent.

95. Following is a truth table for both *not* (p and q) and (*not* p) *and* (*not* q).

p	q	p and q	*not* (p and q)	(*not* p) *and* (*not* q)
T	T	T	F	F
T	F	F	T	F
F	T	F	T	F
F	F	F	T	T

Note that the last two columns in the truth table don't agree, and thus the statements are not logically equivalent.

97. Following is a truth table for (*p and q*) *or r* and (*p or r*) *and* (*p or q*).

p	*q*	*r*	*p and q*	(*p and q*) or *r*	*p or r*	*p or q*	(*p or r*) and (*p or q*)
T	T	T	T	T	T	T	T
T	T	F	T	T	T	T	T
T	F	T	F	T	T	T	T
T	F	F	F	F	T	T	T
F	T	T	F	T	T	T	T
F	T	F	F	F	F	T	F
F	F	T	F	T	T	F	F
F	F	F	F	F	F	F	F

Since the fifth and eighth column of the table don't agree, these two statements are not logically equivalent.

99. Given the implication *if p, then q*, the contrapositive is (*not q*) *then* (*not p*). The converse is *if q, then p* and the inverse of the converse is *if* (*not q*) *then* (*not p*), which is the contrapositive. Similarly, the contrapositive is also the converse of the inverse.

UNIT 1C

QUICK QUIZ

1. **b.** The ellipsis is a convenient way to represent all the other states in the U.S. without having to write them all down.

2. **c.** $3\frac{1}{2}$ is a rational number (a ratio of two integers), but it is not an integer.

3. **a.** When the circle labeled C is contained within the circle labeled D, it indicates that C is a subset of D.

4. **b.** Since the set of boys is disjoint from the set of girls, the two circles should be drawn as non-overlapping circles.

5. **a.** Because all apples are fruit, the set A should be drawn within the set B (the set of apples is a subset of the set of fruits).

6. **c.** Some cross country runners may also be swimmers, so their sets should be overlapping.

7. **a.** The X is placed in the region where *business executives* and *working mothers* overlap to indicate that there is at least one member in that region.

8. **c.** The region X is within both *males* and *athletes*, but not within *Republicans*.

9. **a.** The central region is common to all three sets, and so represents those who are male, Republican, and an athlete.

10. **c.** The sum of the entries in the column labeled Low Birth Weight is 32.

DOES IT MAKE SENSE?

7. Does not make sense. More likely than not, the payments go to two separate companies.

9. Does not make sense. The number of students in a class is a whole number, and whole numbers are not in the set of irrational numbers.

11. Does not make sense. A Venn diagram shows only the relationship between members of sets, but does not have much to say about the truth value of a categorical proposition.

BASIC SKILLS AND CONCEPTS

13. 23 is a natural number.

15. 2/3 is a rational number.

17. 1.2345 is a rational number.

19. π is a real number.

21. −34.45 is a rational number.

23. $\pi/4$ is a real number.

25. −13/3 is a rational number.

27. $\pi/129$ is a real number.

29. {January, February, March, …, November, December}

31. {New Mexico, Oklahoma, Arkansas, Louisiana}

33. {9, 16, 25}

35. {3, 9, 15, 21, 27}

37. Because some men are attorneys, the circles should overlap.

39. Water is a liquid, and thus the set of water is a subset of the set of liquids. This means one circle should be contained within the other.

41. Some novelists are also athletes, so the circles should overlap.

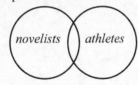

43. No rational number is an irrational number, so these sets are disjoint, and the circles should not overlap.

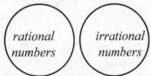

45. b. The subject is *widows*, and the predicate is *women*.
    c.

   d. No, the diagram does not show evidence that there is a woman that is not a widow.

47. a. All U.S. presidents are people over 30 years old.
    b. The subject is *U.S. presidents*, and the predicate is *people over 30 years old*.
    c.

   d. Yes, no U.S. presidents are outside the set of people over 30.

49. a. No monkey is a gambling animal.
    b. The subject is *monkeys*, and the predicate is *gambling animals*.
    c.

   d. No, since the sets are disjoint, the would have no common members.

51. a. All winners are people who smile.
    b. The subject is *winners*, and the predicate is *people who smile*.
    c.

   d. Yes, since all winners are inside the set of people that smile, no frowner can be a winner.

53.

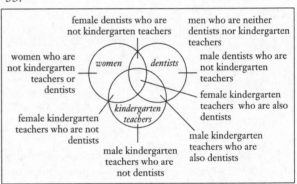

55.

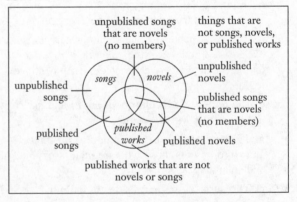

57.

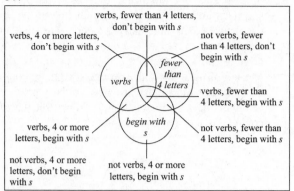

59. a. There are 16 women at the party that are under 30.
b. There are 22 men at the party that are not under 30.
c. There are 44 women at the party.
d. There are 81 people at the party.

61.

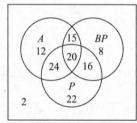

63. a. There are 20 people at the conference that are unemployed women with a college degree.

b. There are 22 people at the conference that are employed men.
c. There are 8 people at the conference that are employed women without a college degree.
d. There are 34 people at the conference that are men.

65. a.

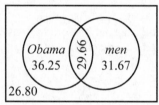

b. Add the numbers in the regions that are contained in the *A* and *BP* circles, to find that 95 people took antibiotics or blood pressure medication.

c. Add the number of people that are in the *BP* circle, but outside the *P* circle, to arrive at 23 people.
d. Add the number of people that are in the *P* circle. There are 82 such people.
e. Use the region that is common to the *A* and *BP* circles, but not contained in the *P* circle, to find

that 15 people took antibiotics and blood pressure medicine, but not pain medication.

f. Add the numbers in the regions that are in at least one of the three circles, to find that 117 people took antibiotics or blood pressure medicine or pain medicine.

FURTHER APPLICATIONS

67. a.

	Favorable Review	Non-favorable Review	Total
Comedy	8	23 – 8 = 15	23
Non-comedy	22 – 12 = 10	12	45 – 23 = 22
Total	8 + 10 = 18	15 + 12 = 27	45

b.

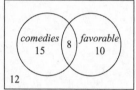

c. 15 comedies received unfavorable reviews.

d. 10 non-comedies received favorable reviews.

69. a.

	Hip-hop	Rock	Total
NY	30	40 – 30 = 10	100 – 60 = 40
LA	20	40	20 + 40 = 60
Total	30 + 20 = 50	10 + 40 = 50	100

b.

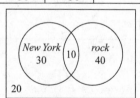

c. 10 New Yorkers preferred rock.

71.

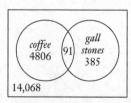

73. The two-way table should look like this:

	Vegetarian	Meat/Fish	Total
Wine	20	40	60
No wine	45	15	60
Total	65	55	120

75.

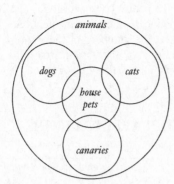

77.

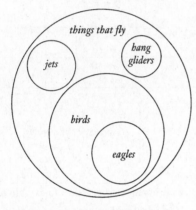

79. a.

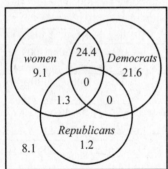

b.

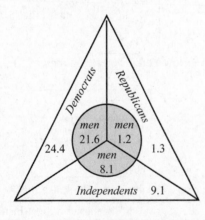

81. Since beans are contained within the set of plants, and meat and dairy products are outside of the set of plants, beans and dairy products are disjoint sets, and thus no bean is a dairy product. There is nothing in the propositions that prohibits the overlap of meat and dairy products, so there could be a meat that is a dairy product. No dairy product is a plant, because it is disjoint from that set. There could be plants that contain protein.

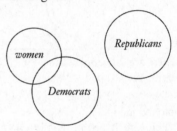

83. a. It is implied here that no Democrat is a Republican, and thus those sets are disjoint, as shown in the diagram.

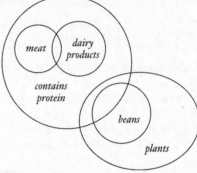

b. Yes, it is possible to meet such a woman, as there is nothing in the propositions to exclude that possibility.

c. Yes, there may be men who are Republicans (if there are any Republicans in attendance, they must be men).

85. a. There are 16 different sets of options: choose nothing, A, B, C, D, AB, AC, AD, BC, BD, CD, ABC, ABD, ACD, BCD, and ABCD.

b.

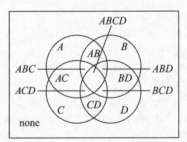

c. There are no regions for A and D only, and for B and C only.

85. (continued)

d. There are 32 different sets of options: choose nothing, A, B, C, D, E, AB, AC, AD, AE, BC, BD, BE, CD, CE, DE, ABC, ABD, ABE, ACD, ACE, ADE, BCD, BCE, BDE, CDE, ABCD, ABCE, ABDE, ACDE, BCDE, and ABCDE.

UNIT 1D

QUICK QUIZ

1. **b**. The only way to prove a statement true beyond all doubt is with a valid and sound deductive argument.

2. **c**. A deductive argument that is valid has a logical structure that implies its conclusion from its premises.

3. **c**. If a deductive argument is not valid, it cannot be sound.

4. **a**. Premise 1 claims the set of *knights* is a subset of the set of *heroes*, and Premise 2 claims Paul is a hero, which means the X must reside within the *hero* circle. However, we cannot be sure whether the X should fall within or outside the *knights* circle, so it belongs on the border.

5. **c**. Diagram *a* in question 4 is the correct diagram for its argument, and since X lies on the border of the *knights* circle, Paul may or may not be a knight.

6. **b**. The argument is of the form *denying the conclusion*, and one can always conclude *p* is not true in such arguments. (Whether the argument is sound is another question).

7. **c**. This argument is of the form *affirming the conclusion*, and it is always invalid, which means we can conclude nothing about *p*.

8. **c**. A chain of conditionals from *a* to *d* is necessary before we can claim the argument is valid.

9. **b**. The side opposite the right angle in a right triangle is always the longest, and it's called the hypotenuse.

10. **b**. The Pythagorean theorem states that $c^2 = 4^2 + 5^2 = 16 + 25 = 41$.

DOES IT MAKE SENSE?

9. Does not make sense. One cannot prove a conclusion beyond all doubt with an inductive argument.

e. Notice that with four options, there are $2^4 = 16$ different sets of options, and that with five options, there are $2^5 = 32$ different sets of options. It turns out that this pattern continues so that with *N* options, there are 2^N different sets of options.

11. Makes sense. As long as the logic of a deductive argument is valid, if one accepts the truth (or soundness) of the premises, the conclusion necessarily follows.

13. Does not make sense. This argument is of the form *affirming the conclusion*, and it is always invalid.

BASIC SKILLS AND CONCEPTS

15. This is an inductive argument because it makes the case for a general conclusion based on many specific observations.

17. This is an inductive argument because it makes the case for a general conclusion based on many specific observations.

19. This is an inductive argument because it makes the case for a general conclusion based on many specific observations.

21. This is an inductive argument because it makes the case for a general conclusion based on several specific observations.

23. The premises are true and the argument is moderately strong. The conclusion is correct.

25. The premises are true, the argument seems moderately strong, and the conclusion is false.

27. The premises are true, and the argument is moderately strong. The conclusion is true.

29. a. Premise: All European countries are countries that use the euro as currency.

b.

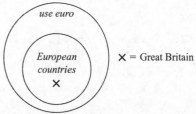

c. The diagram shows the argument is valid. However, it is not sound as the first premise is false.

31. a. All states west of the Mississippi River are not in the eastern time zone.
    b.

states west of Mississippi River ✕   eastern time zone ✕ = Utah

c. The argument is valid, and the premises are true, so the argument is also sound.

33. b. As shown in the diagram, the argument is not valid because we cannot place the X within the *Best Actor Award* circle based on the second premise alone.

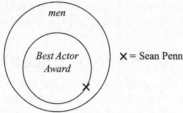

men / Best Actor Award   ✕ = Sean Penn

c. Though the premises are true, the argument is not sound.

35. a. All CEOs are people who can whistle a Springsteen tune.
    b. The argument is valid

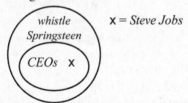

whistle Springsteen / CEOs ✕   ✕ = *Steve Jobs*

c. The premises could be true, in which case the argument is sound.

37. b. Affirming the hypothesis – this form is always valid, as confirmed by the diagram.

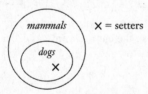

mammals / dogs ✕   ✕ = setters

c. The premises are true, and thus the argument is sound.

39. b. Denying the hypothesis – this form is always invalid, as confirmed by the diagram.

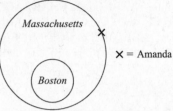

Massachusetts / Boston ✕   ✕ = Amanda

c. Since it is invalid, it cannot be sound.

41. b. Denying the conclusion – this form is always valid, as confirmed by the diagram.

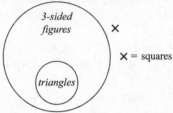

3-sided figures / triangles ✕   ✕ = squares

c. Since the argument is valid and the premises are true, this is a sound argument.

43. a. If a novel was written in the 19th century, then it was not written on a word processor.
    b. Denying the hypothesis – this form is always invalid, as confirmed by the diagram.

novels written in 19th century   ✕ novels written with word processor   ✕ = Jake's novel

c. Since it is invalid, it cannot be sound.

45. The argument is in standard form, and it is valid as there is a clear chain of implications from the first premise to the conclusion.

47. The second premise and conclusion should be written as follows to put the argument in standard form. Premise: If taxpayers have less disposable income, the economy will slow down. Conclusion: If taxes are increased, then the economy will slow down. The argument is valid as there is a clear chain of implications from premises to conclusion.

49. The statement is true.

51. The statement is not true. Counterexamples will vary. One possibility is: $\sqrt{9+16} = \sqrt{25} = 5$; $\sqrt{9} + \sqrt{16} = 3 + 4 = 7$; but $5 \neq 7$.

FURTHER APPLICATIONS

53. It is possible
    Answers will vary. An example:
    Premise: All living mammals breathe.
    Premise: All monkeys are mammals.
    Conclusion: All living monkeys breathe.

55. It is possible
    Answers will vary. An example:
    Premise: All mammals fly. (false)
    Premise: All monkeys are mammals. (true)
    Conclusion: All monkeys fly. (false)

57.   It is possible
    Answers will vary. An example:
    Premise: All mammals breathe. (true)
    Premise: All mammals have hair. (true)
    Conclusion: All hairy animals breathe. (true)

59. An example of *affirming the conclusion* (invalid):
    Premise: If I am in Phoenix, then I am in Arizona.
    Premise: I am in Arizona.
    Conclusion: I am in Phoenix.

61. An example of *denying the conclusion* (valid):
    Premise: If I am in Phoenix, then I am in Arizona.
    Premise: I am not in Arizona.
    Conclusion: I am not in Phoenix.

63. a. Someone has a huge hole in his portfolio.
    b. Lehman Brothers was able to pay out on its losing bets.
    c. No conclusion can be made since this scenario is not covered by the premises.

65. a. That individual expresses righteous indignation.
    b. No obsessive individual got emotional.
    c. No conclusion can be made since this scenario is not covered by the premises.

UNIT 1E

QUICK QUIZ

1. **b.** A vote for C implies a property tax reduction.

2. **c.** An argument that doesn't clearly spell out all of its premises is weak in logical structure.

3. **c.** With unknown application fees, it's not clear which bank has the better offer.

4. **b.** It's a good deal if you get six haircuts at this shop within a year (and that you remember to get your card punched), but it's a bad deal otherwise.

5. **c.** $20/100 min = 20¢/min.

6. **a.** As long as you remember to get the 50% refund coming to you, you'll spend $200.

7. **c.** Both Jack's argument and this one make a huge leap from premises to conclusion: here's a few examples when *A* happened, so *A* must happen all the time.

8. **b.** You can't compute how much you'll spend with each policy without knowing the number and cost of collisions over the span of a year.

9. **a.** The teacher is assuming that students will do fine without spell checkers, which implies that traditional methods of teaching spelling are effective.

10. **c.** If it did not rain, and today is a Saturday, the Smiths would have a picnic. Since they did not, it must not be a Saturday.

DOES IT MAKE SENSE?

5. Makes sense. The double negative means the insurance company accepted his claim.

7. Does not make sense. If Sue wants to save time, she should take the Blue Shuttle, and save ten minutes.

9. Makes sense. Both the duration and mileage of the first warranty is the better deal.

BASIC SKILLS AND CONCEPTS

11. He has 4 bagels left as he ate all but 4.

13. Neither person, roosters don't lay eggs.

15. You must meet 22 people, as the first twenty might all be Canadians.

17. You must meet three people, as the first two might be different nationalities.

19. Suzanne might go bowling 1, 2, 3, or 4 days per week.

21. No, it does not follow. All of the chocolate lovers may be men.

23. a. A person can serve for three consecutive terms of four years, or 12 years.
    b. The councilmember would have to wait 8 years.
    c. No; the councilmember must only wait 8 years if he or she has served for three consecutive terms.
    d. The councilmember could serve for three more consecutive terms since this is not prohibited by the charter.

25. (1) Buying a house will continue to be a good investment. (2) You will spend less out-of-pocket on your home payments than you would on rent.

27. (1) The Governor will keep his promise on tax cuts. (2) you consider tax cuts to be more important than other issues.

29. We are looking for possible unstated motives that may be the unstated "real reason" for opposition to the spending proposal. Among the possibilities is that the speaker may have a fundamental ideological opposition to paying taxes.

31. a. Maria must file a return since her earned income is greater than $5950.
    b. Van must file a return since his gross income of $3500 is greater than his earned income plus $300 ($3500 > $3000 + $300).
    c. Walt need not file a return since his income does not meet any of the criteria.
    d. Helena must to file a return since her gross income of $6000 is greater than her earned income (up to $5650) plus $300 ($6000 > $5650 + $300).

33. a. The landlord has one month after June 5 to return the deposit, so the terms have been met.
    b. The landlord has one month after June 5 to return the deposit, so the terms have been met.
    c. The landlord has one month after June 5 to return the deposit, so the terms have not been met.

35. a. Since you will probably pay for service and insurance with either plan, those costs should not determine which option you choose.
    b. Yes; the total cost of the car at the end of the lease is $1000 + $240 × 36 + $9000 = $18,640, which is greater than the purchase price of $18,000.
    c. If you lease the care, you have years to decide if you want to buy it, don't need to worry about selling the car, and the dealer may offer special servicing prices with a lease. Leasing may also be better if you do not plan to keep the car for a long period of time.

FURTHER APPLICATIONS

Were there more than 350 people on each jet, or both jets combined? Was there any damage or injuries?

39. Is the elephant in your pajamas or were you wearing the pajamas?

41. a. They could be consistent, because Alice does not specify the time period for her 253 cases, and it's possible Zack is being selective about the cases he chooses to discuss (that is, he may not be talking about all the cases Alice tried).
    b. They could be consistent if Alice obtained many convictions through plea agreements without going to trial.

43. a. New conditions go into effect without user approval.
    b. No, continued use of the software implies user acceptance.
    c. New conditions that affect the user could go into effect without user knowledge or approval.
    d. It is impossible to distinguish a typographical error from a deliberate attempt to take advantage of users.

45. Under your current policy (and over the span of a nine-month pregnancy), you'll spend $115 per month, plus $4000 for prenatal care and delivery, for a total of $5035. Under the upgraded policy, you'll spend $275 per month, for a total of $2475. Thus, considering only these costs (we aren't told, for example, what happens if the mother requires an extensive hospital stay due to a C-section or other complications arising from delivery, or what happens if the baby is born prematurely and requires neo-natal care), the upgraded policy is best.

47. If you fly nine times with Airline A, at a cost of $3150, your tenth flight will be free (because you will have flown 27,000 miles at that point). If you fly ten times with Airline B, your cost will be $3250. Thus if you plan to fly ten times (or a multiple of ten times), Airline A is cheaper than Airline B.

49. Winning four of six games is only one more win than the three wins that would be expected by pure chance.

51. People convicted of violet crimes were given longer prison sentences.

53. If the population increases quickly enough, then the death *rate* can decrease even though the number of deaths increases.

55. Country Y has a high gun suicide rate.

57–65. Answers will vary.

UNIT 2A

QUICK QUIZ

1. **a.** Think of the unit *miles per hour*; the unit of *mile* is divided by the unit of *hour*.

2. **b.** The area of a square is its length multiplied by its width (these are, of course, equal for squares), and thus a square of side length 2 mi has area of $2 \text{ mi} \times 2 \text{ mi} = 4 \text{ mi}^2$.

3. **c.** When multiplying quantities that have units, the units are also multiplied, so $\text{ft}^2 \times \text{ft} = \text{ft}^3$.

4. **b.** $1 \text{ mi}^3 = (1760 \text{ yd})^3 = 1760^3 \text{ yd}^3$.

5. **c.** $1 \text{ ft}^2 = 12 \text{ in} \times 12 \text{ in} = 144 \text{ in}^2$.

6. **a.** Divide both sides of 1 L = 1.057 qt by 1 L.

7. **c.** The metric prefix *kilo* means 1000, so a kilometer is 1000 meters.

8. **c.** Water boils at 100°C (at sea level), so 110°C is boiling hot.

9. **a.** Apples are most likely to be sold by units of weight (or more accurately, mass), and thus euros per kilogram is the best answer.

10. **b.** $1.32 per euro means 1 euro = $1.32, which is more than $1.

DOES IT MAKE SENSE?

7. Does not make sense. 35 miles is a distance, not a speed.

9. Makes sense. Liquids are measured in liters, and since one liter is about a quart, drinking two liters is a reasonable thing to do.

11. Makes sense. 10,000 meters is 10 kilometers, which is about 6.2 miles, a common length for foot races. Anyone who can run six back-to-back 9-minute miles has no trouble running 10,000 meters in less than an hour.

BASIC SKILLS AND CONCEPTS

13. a. $\dfrac{3}{4} \times \dfrac{1}{2} = \dfrac{3 \cdot 1}{4 \cdot 2} = \dfrac{3}{8}$

    b. $\dfrac{2}{3} \times \dfrac{3}{5} = \dfrac{2 \cdot 3}{3 \cdot 5} = \dfrac{2}{5}$

    c. $\dfrac{1}{2} + \dfrac{3}{2} = \dfrac{1+3}{2} = \dfrac{4}{2} = 2$

    d. $\dfrac{2}{3} + \dfrac{1}{6} = \dfrac{4}{6} + \dfrac{1}{6} = \dfrac{4+1}{6} = \dfrac{5}{6}$

    e. $\dfrac{2}{3} \times \dfrac{1}{4} = \dfrac{2 \cdot 1}{3 \cdot 4} = \dfrac{2}{12} = \dfrac{1}{6}$

f. $\dfrac{1}{4} + \dfrac{3}{8} = \dfrac{2}{8} + \dfrac{3}{8} = \dfrac{2+3}{8} = \dfrac{5}{8}$

g. $\dfrac{5}{8} - \dfrac{1}{4} = \dfrac{5}{8} - \dfrac{2}{8} = \dfrac{5-2}{8} = \dfrac{3}{8}$

h. $\dfrac{3}{2} \times \dfrac{2}{3} = \dfrac{3 \cdot 2}{3 \cdot 2} = 1$

15. Answers may vary depending on whether fractions are reduced.

    a. $3.5 = \dfrac{35}{10} = \dfrac{7}{2}$ b. $0.3 = \dfrac{3}{10}$

    c. $0.05 = \dfrac{5}{100} = \dfrac{1}{20}$ d. $4.1 = \dfrac{41}{10}$

    e. $2.15 = \dfrac{215}{100} = \dfrac{43}{20}$ f. $0.35 = \dfrac{35}{100} = \dfrac{7}{20}$

    g. $0.98 = \dfrac{98}{100} = \dfrac{49}{50}$ h. $4.01 = \dfrac{401}{100}$

17. a. $\dfrac{1}{4} = 0.25$.   b. $\dfrac{3}{8} = 0.375$

    c. $\dfrac{2}{3} \approx 0.667$ d. $\dfrac{3}{5} = 0.6$

    e. $\dfrac{13}{2} = 6.5$ f. $\dfrac{23}{6} \approx 3.833$

    g. $\dfrac{103}{50} = 2.06$ h. $\dfrac{42}{26} \approx 1.615$

19. a. $10^4 \times 10^7 = 10^{4+7} = 10^{11}$

    b. $10^5 \times 10^{-3} = 10^{5-3} = 10^2$

    c. $10^6 \div 10^2 = 10^{6-2} = 10^4$

    d. $\dfrac{10^8}{10^{-4}} = 10^{8-(-4)} = 10^{12}$

    e. $\dfrac{10^{12}}{10^{-4}} = 10^{12-(-4)} = 10^{16}$

    f. $10^{23} \times 10^{-23} = 10^{23-23} = 10^0 = 1$

    g. $10^4 + 10^2 = 10,000 + 100 = 10,100$

    h. $10^{15} \div 10^{-5} = 10^{15-(-5)} = 10^{20}$

21. $3.5 \text{ lb} \times \dfrac{\$0.90}{1 \text{ lb}} = \$3.15$

23. $6 \text{ months} \times \dfrac{\$3200}{1 \text{ month}} = \$19,200$

25. a. The area of the arena's floor is 200 ft × 150 ft $= 30,000 \text{ ft}^2$, and the volume of the arena is 200 ft × 150 ft × 35 ft $= 1,050,000 \text{ ft}^3$.

25. (continued)

b. The surface area of the pool is 30 yd × 10 yd = 300 yd^2, and the volume of water it holds is 30 yd × 10 yd × 0.3 yd = 90 yd^3.

c. The area of the bed is 25 ft × 8 ft = 200 ft^2, and the volume of soil it holds is 25 ft × 8 ft × 1.5 ft = 300 ft^3.

27. Speed has units of miles per hour, or mi/hr.

29. The cost of carpet has units of dollars per square yard, or $/ yd^2.

31. The price of rice has units of yen per kilogram, or yen/kg.

33. The daily consumption has units of gallons per person, or gal/person.

35. $24 \text{ ft} \times \dfrac{12 \text{ in}}{1 \text{ ft}} = 288 \text{ in}$

37. $25 \text{ min} \times \dfrac{60 \text{ s}}{1 \text{ min}} = 1500 \text{ s}$

39. $2.5 \text{ hr} \times \dfrac{60 \text{ min}}{1 \text{ hr}} \times \dfrac{60 \text{ s}}{1 \text{ min}} = 9000 \text{ s}$

41. $3 \text{ tr} \times \dfrac{365 \text{ day}}{1 \text{ tr}} \times \dfrac{24 \text{ hr}}{1 \text{ day}} = 26{,}280 \text{ hr}$

43. Note that 1 ft = 12 in, and thus $(1 \text{ ft})^2 = (12 \text{ in})^2$, which means $1 \text{ ft}^2 = 144 \text{ in}^2$. This can also be written as $\dfrac{1 \text{ ft}^2}{144 \text{ in}^2} = 1$, or $\dfrac{144 \text{ in}^2}{1 \text{ ft}^2} = 1$.

45. The volume of the sidewalk is 4 ft × 200 ft × 0.5 ft = 400 ft^3. Since 1 yd = 3 ft, we know 1 yd^3 = 27 ft^3, and this can be used to convert to cubic yards.

$$400 \text{ ft}^3 \times \dfrac{1 \text{ yd}^3}{27 \text{ ft}^3} = 14.8 \text{ yd}^3$$

47. Use the fact that 1 yd^3 = 27 ft^3 (see Exercise 45).

$$320 \text{ ft}^3 \times \dfrac{1 \text{ yd}^3}{27 \text{ ft}^3} = 11.9 \text{ yd}^3$$

49. a. $10 \text{ furlongs} \times \dfrac{1 \text{ mi}}{8 \text{ furlongs}} \times \dfrac{5280 \text{ ft}}{1 \text{ mi}} \times \dfrac{1 \text{ yd}}{3 \text{ ft}}$

$\times \dfrac{1 \text{ rod}}{5.5 \text{ yd}} = 400 \text{ rods}$

b. $10 \text{ furlongs} \times \dfrac{1 \text{ mi}}{8 \text{ furlongs}} \times \dfrac{5280 \text{ ft}}{1 \text{ mi}} \times \dfrac{1 \text{ fathom}}{6 \text{ ft}}$

$= 1100 \text{ fathoms}$

51. Convert a cubic foot of water into pounds.

$$1 \text{ ft}^3 \text{ (of water)} \times \dfrac{7.48 \text{ gal}}{1 \text{ ft}^3} \times \dfrac{8.33 \text{ lb}}{1 \text{ gal}} = 62.3 \text{ lb}$$

Now convert that answer into ounces.

$$62.3 \text{ lb} \times \dfrac{16 \text{ oz}}{1 \text{ lb}} = 997 \text{ oz (av)}$$

53. Use the fact that 1 nautical mile = 6076.1 feet to convert knots into mph.

$$46 \text{ knots} = \dfrac{46 \text{ naut. mi}}{\text{hr}} \times \dfrac{6076.1 \text{ ft}}{1 \text{ naut. mi}} \times \dfrac{1 \text{ mi}}{5280 \text{ ft}} =$$

$52.94 \dfrac{\text{mi}}{\text{hr}}$

55. You can tell the factor by which the first unit is larger than the second by dividing the second into the first. Rewrite metric prefixes as powers of 10, and then simplify, as shown below.

$\dfrac{1 \text{ m}}{1 \text{ mm}} = \dfrac{10^0 \text{ m}}{10^{-3} \text{ m}} = \dfrac{10^0}{10^{-3}} = 10^{0-(-3)} = 10^3$. This means a meter is 1000 times as large as a millimeter.

57. See Exercise 55.

$\dfrac{1 \text{ L}}{1 \text{ mL}} = \dfrac{1 \text{ L}}{10^{-3} \text{ L}} = \dfrac{1}{10^{-3}} = 10^3 = 1000$, so a liter is 1000 times larger than a milliliter.

59. See Exercise 55.

$\dfrac{1 \text{ m}^2}{1 \text{ cm}^2} = \dfrac{1 \text{ m}^2}{(10^{-2} \text{ m})^2} = \dfrac{1 \text{ m}^2}{10^{-4} \text{ m}^2} = \dfrac{1}{10^{-4}} = 10^4$, so a square meter is 10,000 times as large as a square centimeter.

61. $22 \text{ lb} \times \dfrac{2.205 \text{ lb}}{1 \text{ kg}} = 48.51 \text{ lb}$. (Note: you'll get an answer of 48.50 lb if you use the conversion 0.4536 kg = 1 lb).

63. $16 \text{ qt} \times \dfrac{1 \text{ L}}{1.057 \text{ qt}} = 15.14 \text{ L}$.

65. $\dfrac{55 \text{ mi}}{\text{hr}} \times \dfrac{1.6093 \text{ km}}{1 \text{ mi}} = 88.51 \dfrac{\text{km}}{\text{hr}}$

67. Cube both sides of the conversion 2.54 cm = 1 in to find the conversion between cubic centimeters and cubic inches.

$$(2.54 \text{ cm})^3 = (1 \text{ in})^3 \Rightarrow 16.387 \text{ cm}^3 = 1 \text{ in}^3$$

67. (continued)

Now use this to complete the problem.

$$300 \text{ in}^3 \times \frac{16.387 \text{ cm}^3}{1 \text{ in}^3} = 4916.12 \text{ cm}^3$$

69. a. Use $C = \dfrac{F-32}{1.8}$. $C = \dfrac{45-32}{1.8} = 7.2°\text{C}$

   b. Use $F = 1.8C + 32$. $F = 1.8(20) + 32 = 68°\text{F}$.

   c. $F = 1.8(-15) + 32 = 5°\text{F}$

   d. $F = 1.8(-30) + 32 = -22°\text{F}$

   e. $C = \dfrac{70-32}{1.8} = 21.1°\text{C}$

71. a. Use $C = K - 273.15$. $C = 50 - 273.15$
   $= -223.15°\text{C}$

   b. $C = 240 - 273.15 = -33.15°\text{C}$.

   c. Use $K = C + 273.15$. $K = 10 + 273.15$
   $= 283.15 \text{ K}$

73. $60 \text{ pounds} \times \dfrac{\$1.624}{1 \text{ pound}} = \$97.44$ or

   $60 \text{ Pounds} \times \dfrac{\$1}{0.6158 \text{ Pound}} = \$97.43$

75. $450 \text{ euros} \times \dfrac{\$1.320}{1 \text{ euro}} = \$594.00$ or

   $450 \text{ euros} \times \dfrac{\$1}{0.7576 \text{ euro}} = \$593.98$

77. $\dfrac{1.5 \text{ euro}}{1 \text{ L}} \times \dfrac{3.785 \text{ L}}{1 \text{ gal}} \times \dfrac{\$1.320}{1 \text{ euro}} = \dfrac{\$7.49}{\text{gal}}$

FURTHER APPLICATIONS

79. a. Convert $27,800,000 per 80 games into dollars per game.

   $$\frac{\$27,800,000}{80 \text{ games}} = \$347,500/\text{game}$$

   b. Convert $347,500 per game into dollars per minute.

   $$\frac{\$347,500}{1 \text{ game}} \times \frac{1 \text{ game}}{48 \text{ min}} \approx \$7240/\text{min}$$

   c. His total hours spent training were
   $\dfrac{40 \text{ hrs}}{1 \text{ game}} \times 80 \text{ games} = 3200$ hours. His total time

   playing was $\dfrac{48 \text{ min}}{1 \text{ game}} \times \dfrac{1 \text{ hr}}{60 \text{ min}} \times 80 \text{ games} = 64 \text{ hr}$.

   The total time related to basketball was 3200 hr + 64 hr = 3264 hr. Convert $27,800,000 per 3264 hours into dollars per hour.

   $$\frac{\$27,800,000}{3264 \text{ hr}} = \$8517/\text{hr}$$

81. a. Convert 16 gigabytes into pages, using the fact that 2000 bytes are need to store 1 page of text.

   $16 \text{ Gbyte} \times \dfrac{1 \text{ billion bytes}}{1 \text{ Gbyte}} \times \dfrac{1 \text{ page}}{2000 \text{ bytes}} =$
   $8,000,000 \text{ pages}$

   b. $8,000,000 \text{ pages} \times \dfrac{1 \text{ book}}{500 \text{ pages}} = 16,000 \text{ books}$

83. $300 \text{ km} \times \dfrac{1 \text{ L}}{26 \text{ km}} \times \dfrac{1.50 \text{ euro}}{1 \text{ L}} \times \dfrac{\$1.320}{1 \text{ euro}} =$
   $\$22.85$

85. $\dfrac{16 \text{ pounds}}{1 \text{ m}^2} \times \left(\dfrac{1 \text{ m}}{1.094 \text{ yd}}\right)^2 \times \dfrac{\$1}{0.6158 \text{ pound}}$
   $= \dfrac{\$21.71}{\text{yd}^2}$

87. The Cullinan diamond weighs 3106 carats, and:

   $3106 \text{ carats} \times \dfrac{0.2 \text{ g}}{1 \text{ carat}} \times \dfrac{1000 \text{ mg}}{1 \text{ g}} = 621,200 \text{ mg}$

   $3106 \text{ carats} \times \dfrac{0.2 \text{ g}}{1 \text{ carat}} \times \dfrac{1 \text{ kg}}{1000 \text{ g}} \times \dfrac{2.205 \text{ lb}}{1 \text{ kg}} = 1.37 \text{ lb}$

89. The Hope diamond weighs 45.52 carats, so:

   $45.52 \text{ carats} \times \dfrac{0.2 \text{ g}}{1 \text{ carat}} = 9.1 \text{ g}$ and

   $9.1 \text{ g} \times \dfrac{1 \text{ kg}}{1000 \text{ g}} \times \dfrac{2.205 \text{ lb}}{1 \text{ kg}} \times \dfrac{16 \text{ oz}}{1 \text{ lb}} = 0.32 \text{ oz}$.

91. 2.2 ounces of 16-karat gold is
   $2.2 \text{ oz} \times \dfrac{16}{24} = 1.47 \text{ oz}$ of pure gold.

93. a. $1 \text{ centiare} \times \dfrac{1 \text{ are}}{100 \text{ centiare}} \times \dfrac{100 \text{ m}^2}{1 \text{ are}} = 1 \text{ m}^2$.

   b. $1 \text{ km}^2 \times \left(\dfrac{1000 \text{ m}}{1 \text{ km}}\right)^2 \times \dfrac{1 \text{ are}}{100 \text{ m}^2} \times \dfrac{1 \text{ ha}}{100 \text{ are}}$
   $= 100 \text{ ha}$.

93. (continued)

c. From the "By the Way" (Page 74), 1 acre = $43,560 \text{ ft}^2$, so

$$1 \text{ ha} \times \frac{100 \text{ are}}{1 \text{ ha}} \times \frac{100 \text{ m}^2}{1 \text{ are}} \times \left(\frac{3.28 \text{ ft}}{100 \text{ m}}\right)^2$$

$$\times \frac{1 \text{ acre}}{43,560 \text{ ft}^2} \approx 2.47 \text{ acres}.$$

d. $\dfrac{10,000 \text{ euro}}{1 \text{ ha}} \times \dfrac{\$1.320}{1 \text{ euro}} \times \dfrac{1 \text{ ha}}{2.47 \text{ acre}}$

$\approx \$5344/\text{acre}$, so $\$5000/\text{acre}$ is the better option.

UNIT 2B

QUICK QUIZ

1. **b**. Speed is described by distance per unit of time, so dividing a distance by a time is the correct choice.

2. **b**. Intuitively, it makes sense to divide a volume (with units of, say, ft^3) by a depth (with units of ft) to arrive at surface area (which has units of ft^2).

3. **a**. Take dollars per gallon and divide it by miles per gallon, and you get

$$\frac{\$}{\text{gal}} \div \frac{\text{mi}}{\text{gal}} = \frac{\$}{\text{gal}} \times \frac{\text{gal}}{\text{mi}} = \frac{\$}{\text{mi}}.$$

4. **b**. A watt is defined to be one joule per second, which is a unit of power, not energy.

5. **c**. Energy used by an appliance is computed by multiplying the power rating by the amount of time the appliance is used, so you need to know how long the light bulb is on in order to compute the energy it uses in that time span.

6. **a**. Multiplying the population density, which has units of $\dfrac{\text{people}}{\text{mi}^2}$, by the area, which has units of mi^2, will result in an answer with units of people.

7. **b**. Concentrations of gases are often stated in parts per million (and the other two answers make no sense).

8. **c**. Multiplying by 30 kg will give the daily dose, which should be divided by three to determine the dose for one eight hour period.

9. **b**. $4 \text{ L} \times \dfrac{0.08 \text{ gm}}{100 \text{ mL}} \times \dfrac{1000 \text{ mL}}{1 \text{ L}} = 0.08 \text{ gm} \times 40,$

$= 3.2 \text{ gm}$

95. $40 \text{ lb} \times \dfrac{1 \text{ kg}}{2.205 \text{ lb}} \approx 18 \text{ kg}$, so the child should drink

$1000 \text{ mL} + (18-10) \text{ kg} \times \dfrac{50 \text{ mL}}{1 \text{ kg}} = 1400 \text{ mL}$ of

fluid, which is $1400 \text{ mL} \times \dfrac{1 \text{ oz}}{29.547 \text{ mL}} \times \dfrac{1 \text{ glass}}{8 \text{ oz}}$

≈ 5.9 glasses or 6 glasses per day.

97. Answers will vary.

TECHNOLOGY EXERCISES

103. Answers will vary.

10. **c**. $100 \text{ mi}^2 \times \dfrac{25 \text{ people}}{1 \text{ mi}^2} = 2500 \text{ people}$ and

$100 \text{ km}^2 \times \dfrac{25 \text{ people}}{1 \text{ km}^2} = 2500 \text{ people}$

DOES IT MAKE SENSE?

5. Does not make sense. Using the familiar formula *distance = rate × time*, one can see that it would be necessary to divide the distance by the rate (or speed) in order to compute the time, not the other way around.

7. Makes sense. There are 4184 joules in a Calorie, and when 10,000,000 joules is converted into Calories, the result is about 2400 Calories, which is a typical caloric intake for an active person.

9. Does not make sense. The volume of a sphere is $V = \frac{4}{3}\pi r^3$, so the volume of a beach ball with a radius of 20 cm (about 8 inches) would be more than $32,000 \text{ cm}^3$. This translates into a mass of 320,000 grams if the density is 10 grams per cm^3, which is 320 kg. The mass of a beach ball isn't anywhere near that large.

11. Makes sense. If the dose is based on weight, doubling the dose makes sense if you double the weight of the subject.

BASIC SKILLS AND CONCEPTS

13. The car is traveling with speed $\dfrac{45 \text{ mi}}{5 \text{ min}}$, and this needs to be converted to mi/hr.

$$\frac{45 \text{ mi}}{5 \text{ min}} \times \frac{60 \text{ min}}{1 \text{ hr}} = 540 \frac{\text{mi}}{\text{hr}}$$

15. Convert 300 gallons to hours by using the conversion 3.2 gallons = 1 minute.

$$300 \text{ gal} \times \frac{1 \text{ min}}{3.2 \text{ gal}} \times \frac{1 \text{ hr}}{60 \text{ min}} = 1.56 \text{ hr}$$

17. Convert 2.5 ounces to dollars by using the conversion 1 oz = $920.

$$2.5 \text{ oz} \times \frac{\$920}{\text{oz}} = \$2300$$

19. First, note that 305 million people is 3050 groups of 100,000 people each. The mortality rate is

$$\frac{565,650 \text{ deaths}}{305,000,000 \text{ people}} \times \frac{305,000,000 \text{ people}}{3050 \text{ groups of } 100,000} =$$
185 deaths/100,000 people .

21. Convert 3 million births per year into births per minute.

$$\frac{3,000,000 \text{ births}}{\text{yr}} \times \frac{1 \text{ yr}}{365 \text{ d}} \times \frac{1 \text{ d}}{24 \text{ hr}} \times \frac{1 \text{ hr}}{60 \text{ min}} =$$
$$5.7 \frac{\text{births}}{\text{min}}$$

23. The cost to drive 250 miles is

$$250 \text{ mi} \times \frac{1 \text{ gal}}{28 \text{ mi}} \times \frac{\$2.90}{1 \text{ gal}} = \$25.89$$

25. Convert one year into hours of sleep.

$$1 \text{ yr} \times \frac{365 \text{ d}}{\text{yr}} \times \frac{8 \text{ hr (of sleep)}}{\text{d}} = 2920 \text{ hr}$$

27. The solution is wrong. It's helpful to include units on your numerical values. This would have signaled your error, as witness:

$$(incorrect) \; 0.11 \text{ lb} \div \frac{\$7.70}{\text{lb}} = \frac{0.11 \text{ lb}^2}{\$7.70} =$$
$$0.014 \frac{\text{lb}^2}{\$}$$

Notice that your solution, when the correct units are added, produces units of square pounds per dollar, which isn't very helpful, and doesn't answer the question. The correct solution is

$$0.11 \text{ lb} \times \frac{\$7.70}{\text{lb}} = \$0.85$$

29. The solution is wrong. Units should be included with all quantities. When dividing by $11, the unit of dollars goes with the 11 into the denominator (as shown below). Also, while it's reasonable to round an answer to the nearest tenth, it's more useful to round to the nearest hundredth in a problem like this, as you'll be comparing the price per pound of the large bag to the price per pound

of the small bag, which is 39¢ per pound. Here's your solution with all units attached, division treated as it should be, and rounded to the hundredth-place:

$$50 \text{ lb} \div \$11 = \frac{50 \text{ lb}}{\$11} = 4.55 \frac{\text{lb}}{\$}.$$

This actually produces some useful information, as you can see a dollar buys 4.55 pounds of potatoes. If you're good with numbers, you can already see this is a better buy than 39¢ per pound, which is roughly 3 pounds for a dollar. But it's better to find the price per pound for the large bag:

$$\$11 \div 50 \text{ lb} = \frac{\$11}{50 \text{ lb}} = 0.22 \frac{\$}{\text{lb}} = 22¢ \text{ per pound}.$$

Now you can compare it to the price per pound of the small bag (39¢/lb) and tell which is the better buy.

31. 6-ounce bottle   14-ounce bottle
$$\frac{\$3.99}{6 \text{ oz}} = \frac{\$0.665}{1 \text{ oz}} \qquad \frac{\$9.49}{14 \text{ oz}} \approx \frac{\$0.678}{1 \text{ oz}}$$
The 6-ounce bottle is the better deal.

33. The 15-gallon tank will cost $\dfrac{\$55.20}{15 \text{ gal}} = \dfrac{\$3.68}{1 \text{ gal}}$ so the $3.60/gal option is the better deal.

35. Convert 2000 miles into gallons by using the conversion 32 miles = 1 gallon.

$$2000 \text{ mi} \times \frac{1 \text{ gal}}{32 \text{ mi}} = 62.5 \text{ gal}$$

Yes; a car that has half the gas mileage would need twice as much gas. Halving the value of the denominator has the same effect as doubling the value of the fraction.

37. a. The driving time when traveling at 55 miles per hour is $2000 \text{ mi} \times \dfrac{1 \text{ hr}}{55 \text{ mi}} = 36.36 \text{ hr}$, while the time at 70 miles per hour is $2000 \text{ mi} \times \dfrac{1 \text{ hr}}{70 \text{ mi}} = 28.57 \text{ hr}$.

b. Your car gets 38 miles to the gallon when driving at 55 mph, so the cost is

$$2000 \text{ mi} \times \frac{1 \text{ gal}}{38 \text{ mi}} \times \frac{\$3.90}{\text{gal}} = \$205.26 .$$

Your car gets 32 miles to the gallon when driving at 70 mph, so the cost is

$$2000 \text{ mi} \times \frac{1 \text{ gal}}{32 \text{ mi}} \times \frac{\$3.90}{\text{gal}} = \$243.75 .$$

39. a. The rise in sea level is found by dividing the volume of water by the Earth's surface area, so

$$\text{sea level rise} = \frac{2.5 \times 10^6 \text{ km}^3}{340 \times 10^6 \text{ km}^2} = 0.007 \text{ km} \qquad \text{or}$$

about 7 meters.

41. Power is the rate at which energy is used, so the power is 100 Calories per mile. Convert this to joules per second, or watts.

$$\frac{100 \text{ Cal}}{1 \text{ mi}} \times \frac{1 \text{ mi}}{10 \text{ min}} \times \frac{1 \text{ min}}{60 \text{ s}} \times \frac{4184 \text{ J}}{1 \text{ Cal}} =$$

$$\frac{418,400 \text{ J}}{600 \text{ s}} = 697 \text{ W}$$

43. In order to compute the cost, we need to find the energy used by the bulbs in kW-hr, and then convert to cents. This can be done in a single chain of unit conversions, beginning with the idea that energy = power × time.

75 watt bulb: One day will cost

$$75 \text{ W} \times \frac{12 \text{ hr}}{1 \text{ day}} \times \frac{1 \text{ kW}}{1000 \text{ W}} \times \frac{\$0.13}{1 \text{ kW-hr}} = \$0.117 .$$

15 watt bulb: One day will cost

$$15 \text{ W} \times \frac{12 \text{ hr}}{1 \text{ day}} \times \frac{1 \text{ kW}}{1000 \text{ W}} \times \frac{\$0.13}{1 \text{ kW-hr}} = \$0.0234$$

You will save $(\$0.117 - \$0.0234) \times 365$ days = $34.16 in one year.

45. The volume of the block is $(3 \text{ cm})^3 = 27 \text{ cm}^3$. Density is mass per unit volume, so the density of the block is $\frac{20 \text{ g}}{27 \text{ cm}^3} = 0.74 \frac{\text{g}}{\text{cm}^3}$. It will float in water because the density of water is $1 \frac{\text{g}}{\text{cm}^3}$.

47. Population density is people per unit area, so the average population density of the U.S. is

$$\frac{306,000,000 \text{ people}}{3,500,000 \text{ mi}^2} = 87 \frac{\text{people}}{\text{mi}^2} .$$

49. The population density for New Jersey is

$$\frac{8,700,000 \text{ people}}{7417 \text{ mi}^2} = 1173 \frac{\text{people}}{\text{mi}^2} .$$

The population density for Alaska is

$$\frac{680,000 \text{ people}}{571,951 \text{ mi}^2} = 1.2 \frac{\text{people}}{\text{mi}^2} ,$$

which is smaller than New Jersey's population density.

51. a. In one week, a 100-pound person would take

$$1 \text{ week} \times \frac{7 \text{ days}}{1 \text{ week}} \times \frac{25 \text{ mg}}{6 \text{ hr}} \times \frac{24 \text{ hr}}{1 \text{ day}} = 700 \text{ mg} , \qquad \text{so}$$

should take $700 \text{ mg} \times \frac{1 \text{ tablet}}{12.5 \text{ mg}} = 56$ tablets.

b. A 100-pound person should take $700 \text{ mg} \times \frac{5 \text{ mL}}{12.5 \text{ mg}} = 280 \text{ mL}$ of liquid Benadryl.

53. a. BAC is usually measured in units of grams of alcohol per 100 milliliters of blood. A woman who drinks two glasses of wine, each with 20 grams of alcohol, has consumed 40 grams of alcohol. If she has 4000 milliliters of blood, her BAC is

$$\frac{40 \text{ g}}{4000 \text{ mL}} = \frac{0.01 \text{ g}}{\text{mL}} \times \frac{100}{100} = \frac{1 \text{ g}}{100 \text{ mL}} .$$

It is fortunate that alcohol is not absorbed immediately, because if it were, the woman would most likely die – a BAC above 0.4 g/mL is typically enough to induce coma or death.

b. If alcohol is eliminated from the body at a rate of 10 grams per hour, then after 3 hours, 30 grams would have been eliminated. This leaves 10 grams in the woman's system, which means her

BAC is $\frac{10 \text{ g}}{4000 \text{ mL}} = \frac{0.0025 \text{ g}}{\text{mL}} \times \frac{100}{100} = \frac{0.25 \text{ g}}{100 \text{ mL}} .$

This is well above the legal limit for driving, so it is not safe to drive. Of course this solution assumes the woman survives 3 hours of lethal levels of alcohol in her body, because we have assumed all the alcohol is absorbed immediately. In reality, the situation is somewhat more complicated.

FURTHER APPLICATIONS

55. a. A metric mile is $1500 \text{ m} \times \frac{3.28 \text{ ft}}{1 \text{ m}} = 4920 \text{ ft}$. A USCS mile is 5280 ft. Since 4920/5280 = 0.932, the metric mile is 93.2% of the USCS mile.

b. Men's mile: Note: (3:43:13 = 223.13 seconds)

$$\frac{1 \text{ mi}}{223.13 \text{ s}} \times \frac{3600 \text{ s}}{1 \text{ hr}} = 16.13 \text{ mi/hr}$$

Men's metric mile: Note: (3:26:00 = 206 seconds)

$$\frac{4920 \text{ ft}}{206 \text{ s}} \times \frac{1 \text{ mi}}{5280 \text{ ft}} \times \frac{3600 \text{ s}}{1 \text{ hr}} = 16.28 \text{ mi/hr}$$

The record holder in the metric mile ran at a faster pace.

55. (continued)

c. Women's mile: Note: (4:12:56 = 252.56 seconds)

$$\frac{1 \text{ mi}}{252.56 \text{ s}} \times \frac{3600 \text{ s}}{1 \text{ hr}} = 14.25 \text{ mi/hr}$$

Women's metric mile: Note: (3:50:46 = 230.46 seconds)

$$\frac{4920 \text{ ft}}{230.46 \text{ s}} \times \frac{1 \text{ mi}}{5280 \text{ ft}} \times \frac{3600 \text{ s}}{1 \text{ hr}} = 14.56 \text{ mi/hr}$$

The record holder in the metric mile ran at a faster pace.

d. This would be true in both cases since, for both men and women, the record holder in the metric mile ran at a faster pace.

57. a. The volume of the bath is 6 ft × 3 ft × 2.5 ft = 45 ft^3. Fill it to the halfway point, and you'll use 22.5 ft^3 of water (half of 45 is 22.5). (Interesting side note: it doesn't matter which of the bathtub's three dimensions you regard as the height – fill it halfway, and it's always 22.5 ft^3 of water). When you take a shower, you use

$$1 \text{ shower} \times \frac{10 \text{ min}}{\text{shower}} \times \frac{1.75 \text{ gal}}{\text{min}} \times \frac{1 \text{ ft}^3}{7.5 \text{ gal}} = 2.33 \text{ ft}^3$$

of water, and thus you use considerably more water when taking a bath.

b. Convert 22.5 ft^3 of water (the water used in a bath) into minutes.

$$22.5 \text{ ft}^3 \times \frac{7.5 \text{ gal}}{1 \text{ ft}^3} \times \frac{1 \text{ min}}{1.75 \text{ gal}} = 96 \text{ min} .$$

c. Plug the drain in the bathtub, and mark the depth to which you would normally fill the tub when taking a bath. Take a shower, and note how long your shower lasts. Step out and towel off, but keep the shower running. When the water reaches your mark (you used a crayon, and not a pencil, right?), note the time it took to get there. You now have a sense of how many showers it takes to use the same amount of water as a bath. For example, suppose your shower took 12 minutes, and it takes a full hour (60 minutes) for the water to reach your mark. That would mean every bath uses as much water as five showers.

59. a. The depth of Lake Victoria is found by dividing the its volume by its surface area, so

$$\text{depth} = \frac{2750 \text{ km}^3}{68700 \text{ km}^2} = 0.03997 \text{ km} \text{ or about 40 meters.}$$

b. The lost volume is

$$68700 \text{ km}^2 \times 10 \text{ ft} \times \frac{0.3048 \text{ m}}{1 \text{ ft}} \times \frac{1 \text{ km}}{1000 \text{ m}}$$
$$= 209.3976 \text{ km}^3 \text{ or about 210 cubic kilometers.}$$

c. The lost volume as a percentage is $\dfrac{210 \text{ km}^3}{2750 \text{ km}^3}$ $= 0.076$ or 7.6%.

61. Convert 1 week into cubic feet of water.

$$1 \text{ wk} \times \frac{7 \text{ d}}{1 \text{ wk}} \times \frac{24 \text{ hr}}{1 \text{ d}} \times \frac{60 \text{ min}}{1 \text{ hr}} \times \frac{60 \text{ s}}{1 \text{ min}} \times$$
$$\frac{25,800 \text{ ft}^3}{\text{s}} = 15,603,840,000 \text{ ft}^3 ,$$

which is about 15.6 billion cubic feet. Since $\dfrac{15,603,840,000}{1,200,000,000,000} = 0.013$, about 1.3% of the water in the reservoir was released.

63. a. The flow rates are $\dfrac{1.5 \text{ L}}{12 \text{ hr}} \times \dfrac{1000 \text{ mL}}{1 \text{ L}} = 125 \dfrac{\text{mL}}{\text{hr}}$

and $125 \dfrac{\text{mL}}{\text{hr}} \times \dfrac{5 \text{ mg}}{100 \text{ L}} = 6.25 \dfrac{\text{mg}}{\text{hr}}$.

b. The rate is $125 \dfrac{\text{mL}}{\text{hr}} \times \dfrac{15 \text{ gtt}}{1 \text{ mL}} = 1875 \dfrac{\text{gtt}}{\text{hr}}$.

c. $12 \text{ hr} \times 6.25 \dfrac{\text{mg}}{\text{hr}} = 75 \text{ mg}$ of dextrose is delivered.

65. a. The flow rate is $\dfrac{10 \text{ mL}}{1 \text{ hr}} \times \dfrac{300 \text{ mg}}{200 \text{ mL}} = 15 \dfrac{\text{mg}}{\text{hr}}$.

b. The infusion should last $60 \text{ mg} \times \dfrac{1 \text{ hr}}{15 \text{ mg}} = 4$ hours.

67. a. $36 \text{ kg} \times \dfrac{50 \text{ mg/kg}}{1 \text{ day}} = 1800 \text{ mg/day}$ and

$\dfrac{1800 \text{ mg}}{1 \text{ day}} \times \dfrac{1 \text{ tablet}}{300 \text{ mg}} = 6 \text{ tablets/day}$, so the patient should take 1 tablet every four hours.

b. From part **a**, you would need 2 tablets or 600 mg in 8 hours, so $600 \text{ mg} \times \dfrac{5 \text{ mL}}{200 \text{ mg}} = 15 \text{ ml}$ of solution in 8 hours, so the rate would be $\dfrac{15 \text{ mL}}{8 \text{ hr}} \times \dfrac{15 \text{ gtt}}{1 \text{ mL}} = 28.13 \text{ gtt/hr}$.

69. a. Convert kilowatt-hours into joules.

$$900 \text{ kW-hr} \times \frac{3,600,000 \text{ J}}{1 \text{ kW-hr}} = 3,240,000,000 \text{ J} .$$

69. (continued)

b. May has 31 days, so the average power is

$$\frac{3,240,000,000 \text{ J}}{31 \text{ d}} \times \frac{1 \text{ d}}{24 \text{ hr}} \times \frac{1 \text{ hr}}{60 \text{ min}} \times \frac{1 \text{min}}{60 \text{ s}} =$$

$$\frac{1210 \text{ J}}{\text{s}} = 1210 \text{ W}.$$

One could also begin with 900 kW-hr per 31 days, and convert that into watts, using 1 watt = 1 joule per second.

c. First, convert joules into liters.

$$3,240,000,000 \text{ J} \times \frac{1 \text{ L}}{12,000,000 \text{ J}} = 270 \text{ L}.$$

Now convert liters into gallons.

$$270 \text{ L} \times \frac{1 \text{ gal}}{3.785 \text{ L}} = 71.33 \text{ gal}.$$

Thus it would take 270 liters = 71.33 gallons of oil to provide the energy shown on the bill (assuming all the energy released by the burning oil could be captured and delivered to your home with no loss).

71. In order to compute the cost, we need to find the energy used by the outdoor spa in kW-hr, and then convert to dollars. This can be done in a single chain of unit conversions, beginning with the idea that energy = power × time (this is true because power is the rate at which energy is consumed; that is, power = energy/time) and assuming a 30-day month.

$$1500 \text{ W} \times 4 \text{ months} \times \frac{1 \text{ kW}}{1000 \text{ W}} \times \frac{30 \text{ d}}{1 \text{ month}} \times$$

$$\frac{24 \text{ hr}}{1 \text{ d}} \times \frac{\$0.10}{1 \text{ kW-hr}} = \$432$$

73. Since energy = power × time, the power plant can produce (assuming a 30-day month)

$$1.5 \times 10^9 \text{ W} \times 1 \text{ month} \times \frac{30 \text{ d}}{\text{month}} \times \frac{24 \text{ hr}}{\text{d}} \times \frac{1 \text{ kW}}{1000 \text{ W}}$$

$$= 1.08 \times 10^9 \text{ kW-hr} = 1.08 \text{ billion kW-hr}$$

of energy every month. The plant needs

$$1.08 \times 10^9 \ \frac{\text{kW-hr}}{1 \text{ month}} \times \frac{1 \text{ kg}}{450 \text{ kW-hr}} =$$

$$2.4 \times 10^6 \ \frac{\text{kg}}{\text{month}}, \text{ or 2.4 million kg of coal every}$$

month. Because $1.08 \times 10^9 \text{ kW-hr} \times$

$$\frac{1 \text{ home}}{1000 \text{ kW-hr}} = 1.08 \times 10^6 \text{ homes},$$ the plant can serve 1.08 million homes.

75. From Exercise 74, each solar panel covers one square meter. You would need

$$1 \text{ kW} \times \frac{1000 \text{ W}}{1 \text{ kW}} \times \frac{1 \text{ panel}}{50 \text{ W}} = 20 \text{ panels},$$

which is 20 square meters of solar panels.

77. a. Energy = power × time, so the energy capacity of the wind farms over the span of a year is

$$2.5 \times 10^9 \text{ W} \times 1 \text{ yr} \times \frac{365 \text{ d}}{1 \text{ yr}} \times \frac{24 \text{ hr}}{1 \text{ d}} \times \frac{1 \text{ kW}}{1000 \text{ W}} \times 0.30$$

$$= 2.19 \times 10^{10} \text{ kW-hr}$$

Since the wind farms produce 30% of their capacity, on average, the energy produced is

$$= 2.19 \times 10^{10} \text{ kW-hr} \times 0.30 = 6.57 \times 10^9 \text{ kW-hr or}$$
$$6,570,000,000 \text{ kW-hr}.$$

This is enough energy to serve

$$6.57 \times 10^9 \text{ kW-hr} \times \frac{1 \text{ home}}{10,000 \text{ kW-hr}}$$

$$= 657,000 \text{ homes}.$$

b. If the 2.19×10^{10} kW-hr of energy produced by wind farms were instead produced from fossil fuels, there would be

$$6.57 \times 10^9 \text{ kW-hr} \times \frac{1.5 \text{ lb}}{1 \text{ kw-hr}} = 9.855 \times 10^9 \text{ lb},$$

or about 9,855,000,000 pounds of carbon dioxide entering the atmosphere each year.

79. The stand would contain 24 hectares

$$\times \frac{10,000 \text{ m}^2}{1 \text{ hectare}} \times \frac{1 \text{ tree}}{20 \text{ m}^2} \times \frac{400 \text{ fbm}}{1 \text{ tree}} = 4,800,000 \text{ fbm}.$$

So there would be $0.10 \times 4,800,000$ fbm $= 480,000$ fbm if one-tenth of the trees were cut.

Note: Exercise 93 in Section 2A shows the calculation of $\dfrac{10,000 \text{ m}^2}{1 \text{ hectare}} = 1.$

UNIT 2C

QUICK QUIZ

1. **c.** Look at example 1 (*Box Office Receipts*) in this unit.

2. **a.** You must have done something wrong as gas mileage is reported in units of miles per gallon.

3. **b.** Common experience tells us that batteries can power a flashlight for many hours, not just a few minutes nor several years.

4. **b.** An elevator that carried only 10 kg couldn't accommodate even one person (a 150 lb person weighs about 75 kg), and hotel elevators aren't designed to carry hundreds of people (you'd need at least 100 people to reach 10,000 kg).

5. **b.** Refer to the discussion of Zeno's paradox in the text. There, it was shown that the sum of an infinite number of ever-smaller fraction is equal to 2, and thus we can eliminate answers **a** and **c**. This leaves **b**.

6. **a.** If you cut the cylinder along its length, and lay it flat, it will form a rectangle with width equal to the circumference of the cylinder (i.e. 10 in), and with length equal to the length of the cylinder.

7. **c.** A widow is a woman who survives the death of her husband, and thus the man is dead, making it impossible for him to marry anyone.

8. **c.** It could happen that the first 20 balls selected are odd, in which case the next two would have to be even, and this is the first time one can be certain of selecting two even balls.

9. **b.** The most likely explanation is that the A train always arrives 10 minutes after the B train. Suppose the A train arrives on the hour (12:00, 1:00, 2:00,…), while the B train arrives ten minutes before the hour (11:50, 12:50, 1:50,…). If Karen gets to the station in the first 50 minutes of the hour, she'll take the B train; otherwise she'll take the A train. Since an hour is 60 minutes long, 5/6 of the time, Karen will take the B train to the beach. Note that 5/6 of 30 days is 25 days, which is the number of times Karen went to the beach. The other scenarios *could* happen, but they aren't nearly as likely as the scenario in answer **b**.

10. **b.** Label the hamburgers as A, B, and C. Put burgers A and B on the grill. After 5 minutes, turn burger A, take burger B and put it on a plate, and put burger C on the grill. After 10 minutes, burger A is cooked, while burgers B and C are half-cooked. Finish off burgers B and C in the final 5 minutes, and you've cooked all three in 15 minutes.

DOES IT MAKE SENSE?

3. Does not make sense. There is no problem-solving recipe that can be applied to all problems.

5. Does not make sense. The four-step method is applicable to a wide variety of problems, and is not limited to mathematical problems. (It's not a cure-all, either, but it's quite useful, and worth learning).

BASIC SKILLS AND CONCEPTS

7. You won't be able to determine the exact number of cars and buses that passed through the toll booth, but the method of trial-and-error leads to the following possible solutions: (16 cars, 0 buses), (13 cars, 2 buses), (10 cars, 4 buses), (7 cars, 6 buses), (4 cars, 8 buses), (1 car, 10 buses). Along the way, you may have noticed that the number of buses must be an even number, because when it is odd, the remaining money cannot be divided evenly into $2 (car) tolls.

9. a. Based on the first race data, Jordan runs 200 meters in the time it takes Amari to run 190 meters. In the second race, Jordan will catch up to Amari 10 meters from the finish line (because Jordan has covered 200 meters at that point, and Amari has covered 190 meters). Jordan will win the race in the last ten meters, because Jordan runs faster than Amari.
b. Jordan will be 5 meters from the finish line when Amari is 10 meters from the finish line (because Jordan has covered 200 meters at that point, and Amari has covered 190 meters). Jordan will win the race in the because Jordan runs faster than Amari and is in the lead.
c. Jordan will be 15 meters from the finish line when Amari is 10 meters from the finish line (because Jordan has covered 200 meters at that point, and Amari has covered 190 meters). In the time Amari runs 10 meters, Jordan will only run

$$10 \text{ m}_{\text{Amari}} \times \frac{200 \text{ m}_{\text{Jordan}}}{190 \text{ m}_{\text{Amari}}} = 10.53 \text{ m}_{\text{Jordan}}, \qquad \text{so}$$

Amari wins the race.
d. From part **c**, Jordan must start 10.53 meters behind the starting line, since that is how far she would run in the same time Amari finishes the last 10 meters of the race.

11. On the second transfer, there are four possibilities to consider. A) All three marbles are black. B) Two are black, one is white. C) One is black, two are white. D) All three are white. In case A), after the transfer, there are no black marbles in the white pile, and no white marbles in the black pile. In case B), one white marble is transferred to the black pile, and one black marble is left in the white pile. In case C), two white marbles are transferred to the black pile, and two black marbles are left in the white pile. In case D), three white marbles are transferred to the black pile, and three black marbles are left in the white pile. In all four cases, there are as many white marbles in the black pile as there are black marbles in the white pile after the second transfer.

13. Proceeding as in example 6, the circular arc formed by the bowed track can be approximated by two congruent right triangles. Since 1 km is 100,000 cm, the base of one of the triangles is 50,000 cm, and its hypotenuse is 50,005 cm. The height of the triangle can be computed as

$$h = \sqrt{50,005^2 - 50,000^2} = 707 \text{ cm},$$

which is about 7.1 meters.

15. Yes, imagine that at the same time the monk leaves the monastery to walk up the mountain, his twin brother leaves the temple and walks down the mountain. Clearly, the two must pass each other somewhere along the path.

FURTHER APPLICATIONS

17. Note that at least one truck must have passed over the counter as there are an odd number of counts, and we must have an odd number of trucks for the same reason. Beginning with one truck, followed by 3, 5, 7, etc. trucks, and computing the number of cars that go along with these possible truck solutions, we arrive at the following answers: (1 truck, 16 cars), (3 t, 13 c), (5 t, 10 c), (7 t, 7 c), (9 t, 4 c), and (11 t, 1 c).

19. a. The time spent running is $2 \text{ mi} \div 4\dfrac{\text{mi}}{\text{hr}}$

$= 2 \text{ mi} \times \dfrac{1 \text{ hr}}{4 \text{ mi}} = \dfrac{1}{2} \text{ hr}.$

b. The time spent walking is $2 \text{ mi} \div 2\dfrac{\text{mi}}{\text{hr}}$

$= 2 \text{ mi} \times \dfrac{1 \text{ hr}}{2 \text{ mi}} = 1 \text{ hr}.$

c. Not true; more time is spent walking than running.

d. The average speed for the trip is $4 \text{ mi} \div \dfrac{3}{2}\text{hr}$

$4 \text{ mi} \times \dfrac{2}{3 \text{ hr}} = \dfrac{8}{3} \text{ mi/hr}$ or about 2.7 mi/hr.

21. Reuben's birthday is December 31, and he's talking to you on January 1. Thus two days ago, he was 20 years old on December 30. He turned 21 the next day. "Later *next* year," refers to almost two years later, when he will turn 23 (*this* year on December 31, he will turn 22).

23. That man is the son of the person speaking (who, in turn, is the son of his father, and an only child).

25. The woman gained $100 in each transaction, so she gained $200 overall.

27. Select one ball from the first barrel, two balls from the second barrel, three from the third, and so on, ending with ten balls chosen from the tenth barrel. Find the weight of all 55 balls thus chosen. If all the balls weighed one ounce, the weight would be 55 ounces. But we know one of the barrels contains two-ounce balls. Suppose the first barrel contained the two-ounce balls – then the weighing would reveal a result of 56 ounces, and we'd know the first barrel was the one that contains two-ounce balls. If, in fact, the second barrel contained the two-ounce balls, the combined weight would be 57 ounces. A combined weight of 58 ounces means the third barrel contained the heavy balls. Continuing in this fashion, we see that a combined weight of n ounces more than 55 ounces corresponds to the nth barrel containing the two-ounce balls.

29. Turn both the 7-minute and 4-minute hourglasses over to begin timing. At 4 minutes, turn the 4-minute hourglass over again. At 7 minutes, turn the 7-minute hourglass over; there will be one minute of sand left in the 4-minute hourglass. At 8 minutes, the upper chamber of the 4-minute hourglass will be empty, and one minute of sand will have drained into the lower chamber of the 7-minute hourglass. Turn the 7-minute hourglass over to time the last minute. This is only one solution of many. Another: Turn both hourglasses over whenever their upper chambers are empty. After 12 minutes (three cycles of the 4-minute hourglass), there will be 5 minutes of sand in the bottom chamber of the 7-minute hourglass. Turn that hourglass on its side. Now you are prepared to measure a 9-minute interval whenever you please (use the 5 minutes of sand in the 7-minute hourglass, followed by one cycle of the 4-minute hourglass).

31. A balance scale is one that displays a needle in the middle when both sides of the scale are loaded with equal weights. Put six coins on each side of the scale. The side with the heavy coin will be lower, so discard the other six. Now put three coins of the remaining six on each side of the scale, and as before, discard the three that are in the light group. Finally, select two of the remaining three coins, and put one on each side of the scale. The scale will be even if the heavy coin is not on the scale. The scale will tip to one side if the heavy coin was in the final selection. This is just one solution – another begins by dividing the 12 coins into three groups of four coins. See if you can supply the logic necessary to find the heavy coin in that scenario.

33. Since the gray, orange, and pink books are consecutive and the pink book cannot be rightmost, the only possible orders are *gray, orange, pink, other, other* or *other, gray, orange, pink, other*. The only way to add the brown and gold books so that the gold book is not leftmost and there are two books between the brown and gold books is the order *brown, gray, orange, pink, gold*.

35. The visitor should patronize the barber with unkempt hair, because he's the barber that cuts the hair of the other barber (who has a splendid hair cut). Of course we are assuming that these barbers don't travel out of town to get their hair done.

37. When a clock chimes five times, there are four pauses between the chimes. Since it takes five seconds to chime five times, each pause lasts 1.25 seconds. There are nine pauses that need to be accounted for when the clock chimes 10 times, and thus it takes $9 \times 1.25 = 11.25$ seconds to chime 10 times. This solution assumes it takes no time for the clock to chime once (we'd need to know the duration of the chime to solve it otherwise).

39. Each guest has, in fact, spent $9. But not all of the $27 spent is resting in the cash register at the front desk. The front desk has $25, the bellhop has $2 (this adds up to the $27 spent), and each guest has $1, for a total of $30. The reason the problem is perplexing is that we think of the $2 held by the bellhop as a *positive* $2, and so it should be added to the $27 spent by the guests to somehow produce the original $30. But from the perspective of the desk clerk running the cash register, the $2 is a *negative* value – it represents money out of the till. $30 (*positive*) came into the till, $2 (*negative*) went to the bellhop, and $1 (*negative*) went to each of the guests. Thus the equation we should be looking at is 30 − 2 − 1 − 1 − 1 = $25 in the cash register (which is correct), or $27 − $2 = $25 (also correct), but *not* $27 + $2 (which is just silly).

UNIT 3A

QUICK QUIZ

1. **b**. A quantity triples when it is increased by 200%.

2. **b**. The absolute change is 75,000 – 50,000 = 25,000, and the relative change is
$$\frac{75,000 - 50,000}{50,000} = 0.5 = 50\% .$$

3. **c**. A negative relative change corresponds to a decreasing quantity.

4. **c**. Suppose Joshua scored 1000 on the SAT. Then Emily would have scored 1500 (50% more than 1000), and Joshua's score is clearly two-thirds of Emily's score (1000/1500 = 2/3).

5. **c**. 120% of $10 is $12.

6. **c**. Multiplying a pre-tax price by 1.09 is equivalent to adding 9% tax to the price, and thus dividing a post-tax price by 1.09 will result in the original price. To find out how much tax was paid, take the post-tax price, and subtract the original price:
$$\$47.96 - \frac{\$47.96}{1.09} .$$

7. **a**. The relative change in interest rates that go from 6% to 9% is $\frac{9\% - 6\%}{6\%} = 0.5 = 50\%$.

8. **c**. When a price is decreased by 100%, it drops to zero, as the entire amount of the price is deducted from the original price.

9. **c**. 10% of $1000 is $100, so your monthly earnings after one year will be $1100. Each year thereafter, your earnings will increase by more than $100 per year (because 10% of $1100 is more than $100), and the end result will be more than $1500.

10. **b**. The only thing we can conclude with certainty is that she won between 20% and 30% of her races. To prove this rigorously, one could detour into an algebra calculation (let x = number of races entered in high school, let y = number of races entered in college, and show that $\frac{0.3x + 0.2y}{x + y}$ is between 20% and 30%). Try different values for x and y, and compute the win percentage of all the races Emily entered to see that answers **a** and **c** are not correct.

DOES IT MAKE SENSE?

7. Makes sense. If a country's population is declining, then the percent change will be negative (and several of Europe's countries have, indeed, experienced recent declines in population).

9. Makes sense. An older child could easily weigh 25% more than a younger child (think of 100-pound and 125-pound children).

11. Does not make sense. Consider the case where I earn $100, and you earn $120. You earn $20 more than I do, which corresponds to 20% more than me (because 20% of $100 is $20). Even though I earn $20 less than you, I do not earn 20% less than you (because 20% of $120 is not $20). In fact, I earn 1/6 less than you do, which is about 16.7% less.

13. Makes sense. This just means that the incidence of cancer for children living near toxic landfills was seven times larger than the incidence of cancer among other children.

15. Makes sense. If the rate of return on the fund used to be 10%, then a 50% increase would result in a rate of return of 15%.

BASIC SKILLS AND CONCEPTS

17. 2/5 = 0.4 = 40%

19. 0.20 = 1/5 = 20%

21. 150% = 1.5 = 3/2

23. 4/9 = 0.444… = 44.44…%

25. 5/8 = 0.625 = 62.5%

27. 69% = 0.69 = 69/100

29. 7/5 = 1.4 = 140%

31. 4/3 = 1.333… = 133.33…%

33. 8 to 4 = 2. 4 to 8 = 1/2. 8 is 200% of 4. Note that the ratio of B to A is just the reciprocal of the ratio of A to B, and that 200% is simply the ratio of A to B expressed as a percent.

35. 150 to 400 = 3/8. 400 to 150 = 8/3. 150 is 37.5% of 400.

37. 52,252 to 17,774 = 2.94. 17,774 to 52,252 = 0.34. 52,252 is 294% of 17,774.

39. 1.5 million to 2.1 million = 0.71. 2.1 million to 1.5 million = 1.4. 1.5 million is 71 % of 2.1 million.

41. 1.82 million to 2.12 million = 0.86. 2.12 million to 1.82 million = 1.16. 1.82 million is 86% of 2.12 million.

43. 28 is 53.8% of 52 because 28/52 = 0.538.

45. 42,800 is 123% of 34,700 because 42,800/34,700 = 1.23.

47. 1189 is 1350% of 88.1 because 1189/88.1 = 13.50.

49. Clint's absolute change in salary was $28,000 − $20,000 = $8000, and the relative change was $\dfrac{\$28,000 - \$20,000}{\$20,000} = 0.4 = 40\%$. Helen's absolute change was $35,000 − $25,000 = $10,000, and the relative change was $\dfrac{\$35,000 - \$25,000}{\$25,000} = 0.4 = 40\%$. Thus Helen's salary grew more in absolute terms, but the relative change was the same. This happened because we are comparing Helen's larger absolute change to her also larger initial salary.

51. The absolute change is $152,000 − $301,000 = −$149,000. The relative change is $\dfrac{\$152,000 - \$301,000}{\$301,000} = -0.495 = -49.5\%$.

53. The absolute change is 1382 papers − 2226 papers = −844 papers. The relative change is $\dfrac{1382 - 2226}{2226} = -0.379 = -37.9\%$.

55. The relative difference is $\dfrac{266 - 220}{220} = 0.209 = 20.9\%$, which means the gestation period of humans (266 days) is 20.9 percent longer than the gestation period of grizzly bears (220 days).

57. The relative difference is $\dfrac{4200 - 2800}{4200} = 0.333 = 33.3\%$, which means the main span of the Tacoma Narrows bridge (2800 feet) is 33.3 percent shorter than the main span of the Golden Gate bridge (4200 feet).

59. The relative difference is $\dfrac{39,000 - 26,100}{26,100} = 0.494 = 49.4\%$, which means the number of deaths due to poisoning in the United States in a year (39,000) is 49.9 percent greater than the number of deaths due to falls (26,100).

61. Will's height is 122% of Wanda's height because 22% *more than* expresses a relative difference in heights, and this is equivalent to 100% + 22% *of* the reference height (i.e. Wanda's height).

63. Virginia's population is 82% of Georgia's population because 18% *less than* expresses a relative difference in population, and this is equivalent to 100% − 18% *of* the reference population (i.e. Georgia's population).

65. Since the wholesale price is 40% *less than* the retail price, it is 60% *of* the retail price, which means it is 0.6 times the retail price.

67. Since the retail cost is 30% *more than* the wholesale cost, it is 130% *of* the wholesale cost, which means it is 1.3 times the wholesale cost.

69. The absolute change is 2.8% − 2.3% = 0.5 percentage point. The relative change is $\dfrac{2.8 - 2.3}{2.3} = 0.217 = 21.7\%$.

71. The absolute change is 83% − 67% = 16 percentage points. The relative change is $\dfrac{83 - 67}{67} = 0.239 = 23.9\%$.

73. In the first case, 33% of city employees in Freetown ride the bus to work because 10% *more than* 30% is 33%. In the second case, 40% of the city employees in Freetown ride the bus, because 40% is 10 percentage points more than 30%.

75. The final cost is $760 + 0.076 × $760 = $817.76.

77. $\dfrac{91\%}{1.137} = 80\%$ of households had cordless phones in 2000.

79. False. If the national economy began at E, and it shrinks at 4% annually, it would be at $E \times 0.96 \times 0.96 \times 0.96 = 0.885E$, which is 11.5% less than the original value.

81. False. Assign an arbitrary number to your profits before the increases, such as $100. If your profits increase by 4% in the first year, they will be $104. If they decrease by 2% in the next year, they will be 0.98 × $104 = $101.92. This is an increase of 1.92%, not 2%.

83. Not possible. A bill can decrease at most by 100%. Once you've deducted 100% from a bill, there's nothing more to pay.

85. Possible. An increase of 100% means restaurant prices are twice what they used to be.

87. Possible. This means your computer is five times faster than my computer.

89. No, the mean score for both classes is not 85%. Assuming a 100-point exam, the first class scored a total of 25 × 86 = 2150 points, and the second class scored a total of 30 × 84 = 2520 points. Thus the combined number of points scored by 55 students is 2150 + 2520 = 4670, and the mean score for both classes is 4670/55 = 84.9%.

FURTHER APPLICATIONS

91. True. 10% of 60% is 1/10 of 60%, which is 6%.

93. False. Some of the hotels may have both a pool and a restaurant, and those would be counted twice if the percentages were just added together.

95. The total number of undergraduates is 2410/0.54 = 4463.

97. The cost for the dinner before the tip was $76.40/1.20 = $63.67.

99. Simon's gross pay is $2200/0.79 = $2785.

101. Speculative credit default swaps are worth $62 trillion × 0.80 = $49.6 trillion.

103. The relative change is $\frac{8.5-7.1}{7.1} = 0.197 = 19.7\%$.

105. They paid $\frac{\$1,350,000}{1.2} = \$1,125,000$ or $1.125 million in 2007.

107. $1.093 \times 1.088 \times 1.089 \times 1.093 \times 1.05 \times 1.087 = 1.61$, so tuition increased by 61%.

UNIT 3B

QUICK QUIZ

1. **a.** The decimal point must move seven places to the left to transform 70,000,000 into scientific notation, and thus its value is 7×10^7 .

2. **c.** 1277 is roughly 10^3, and 14,385 is of order 10^4, so their product is around 10^7 .

3. **c.** 10^9 is larger than 10^5 by a factor of 10^4, which is ten thousand.

4. **c.** One person, who drives 15,000 miles in a year in a vehicle that gets 25 mpg, uses 600 gallons of gas per year. There are 300 million Americans, and if each used this much gas, the total would be

   3×10^8 people $\times \dfrac{6 \times 10^2 \text{ gal}}{\text{person}} = 18 \times 10^{10}$ gal =

   1.8×10^{11} gallons of gasoline, which is 180 billion gallons. Of course, not everyone drives 15,000 miles every year (children, the elderly, and urban dwellers come to mind). On the other hand, this estimate doesn't even consider all the gas used by trucking, airlines, and other industries. Certainly the answer can't be **a** or **b**, as these estimates are much too small. 100 billion gallons is probably a conservative estimate.

5. **c.** A dollar bill is about 6 inches long, or about 15 cm long, and thus 8 million of them amounts to

   8×10^6 bills $\times \dfrac{1.5 \times 10^1 \text{ cm}}{\text{bill}} \times \dfrac{1 \text{ km}}{10^5 \text{ cm}} =$

   12×10^2 km, which is 1200 kilometers. This is enough to get into outer space, but not near enough to reach the moon.

6. **c.** If you know the song from the Broadway musical *Rent*, you may recall that there are 525,600 minutes in a year. 1.2×10^{-10} year is just over one ten-billionth of a year, which isn't even close to a minute (it's a small fraction of a second). Clearly the sun won't burn out in the next fraction of a second, nor in the next minute.

7. **b.** If 1 inch = 20 miles, then 6 inches is 6 × 20 = 120 miles.

8. **b.** $1 billion is $1000 million, so at $10 million per year, it will take 100 years to earn $1 billion.

9. **c.** Assuming you're willing to campaign for eight hours per day (which is roughly 500 minutes), and that you'll limit yourself to 10 minutes of talking at each house (including travel time between houses), you'll be able to visit 50 households per day, or 100 houses in two days. This translates into 500 houses in two weeks (campaigning five days per week), or 1000 households in a month, or 12,000 households in a year-long campaign, where you do nothing but campaigning. Clearly it would be impossible to carry out your plan before the election – pay for some TV advertisements instead.

10. **a.** The largest Division 1 college football stadiums hold about 100,000 fans. In a lottery with the stated odds, there's only one winning ticket for every 1,000,000 printed, which means there are 999,999 losing tickets. More likely than not, the fans in the stadium will be disappointed (and it's more likely still they will be disappointed in a smaller stadium).

DOES IT MAKE SENSE?

9. Makes sense. The page you are reading right now has two columns, each of which has about 50 lines, and there are roughly 10 words per line, for a total of 1000 words per page. A book with 10^5 = 100,000 words would have 100 pages. A paperback with pages about half the size of this solutions manual would be two-hundred pages long.

11. Makes sense. At ten feet per floor, an apartment building with 20 floors would be 200 feet tall.

13. Does not make sense. A typical NFL stadium holds 70,000 fans. At an unrealistic speed of 10 signatures per minute, it would take the football star 7000 minutes (roughly 100 hours) to provide signatures for all the fans (who would grow restless waiting in line for a signature after about 15 minutes).

BASIC SKILLS AND CONCEPTS

15. a. $3 \times 10^3 = 3000 =$ three thousand

    b. $6 \times 10^6 = 6,000,000 =$ six million

    c. $3.4 \times 10^5 = 340,000 =$ three hundred forty thousand

    d. $2 \times 10^{-2} = 0.02 =$ two hundredths

    e. $2.1 \times 10^{-4} = 0.00021 =$ twenty-one hundred-thousandths

    f. $4 \times 10^{-5} = 0.00004 =$ four hundred - thousandths

17. a. $233 = 2.33 \times 10^2$

    b. $126,547 = 1.26547 \times 10^5$

    c. $0.11 = 1.1 \times 10^{-1}$

    d. $9736.23 = 9.73623 \times 10^3$

    e. $124.58 = 1.2458 \times 10^2$

    f. $0.8642 = 8.642 \times 10^{-1}$

19. a. $(3 \times 10^3) \times (2 \times 10^2) = 6 \times 10^5$

    b. $(4 \times 10^2) \times (3 \times 10^8) = 12 \times 10^{10} = 1.2 \times 10^{11}$

    c. $(3 \times 10^3) + (2 \times 10^2) = 3000 + 200 = 3200$
    $$= 3.2 \times 10^3$$

    d. $(8 \times 10^{12}) \div (4 \times 10^4) = 2 \times 10^8$

21. a. 10^{35} is 10^9 (1 billion) times as large as 10^{26}.

    b. 10^{27} is 10^{10} (10 billion) times as large as 10^{17}.

    c. 1 billion is 10^3 (1000) times as large as 1 million.

23. My new music player has a capacity of 3.4×10^{11} bytes.

25. The diameter of a typical bacterium is about 1×10^{-6} meter.

27. a. 300,000 is 3×10^5, and 100 is 1×10^2, so their product is 3×10^7, or 30,000,000. This is an exact answer as no estimation was necessary.

    b. 5.1 million is about 5×10^6, and 1.9 thousand is about 2×10^3, so their product is about 1×10^{10}, which is 10 billion. The exact answer (computed with a calculator) is 9.69 billion.

    c. Since 2.1×10^6 is about 2 million, $4 \times 10^9 \div 2.1 \times 10^6$ is about 2×10^3, or 2000. The exact answer is nearer to 1.9×10^3 (1904.76 is a rounded answer).

29. Answers will vary regarding the estimates made, but most people spend more money on gas each month than on coffee. If you buy, on average, one coffee per day, spending $2 per coffee, that's around $60 per month. On the other hand, if you fill up your gas tank once a week, spending around $50 to fill a tank (assuming a 12 gallon tank and $4.00/gallon for gas,) that's around $200 to $250 per month.

31. Yes, assuming a dime weighs 2.27 g, the weight of $100 in dimes would be

    $$\$100 \times \frac{10 \text{ dimes}}{\$1} \times \frac{2.27 \text{ g}}{1 \text{ dime}} \times \frac{1 \text{ oz}}{28.35 \text{ g}} \times \frac{1 \text{ lb}}{16 \text{ oz}} = 5 \text{ lb}.$$

33. Assuming 1 beat per second, one can estimate the number of heart beats in a day with a unit conversion:

    $$1 \text{ d} \times \frac{24 \text{ hr}}{\text{d}} \times \frac{60 \text{ min}}{\text{hr}} \times \frac{60 \text{ sec}}{\text{min}} \times \frac{1 \text{ beat}}{\text{sec}}, \text{ which is}$$
    about 86,400 beats per day (answers will vary).

35. Answers will vary, but assuming a daily water intake of 2 pints, one can estimate the annual water consumption with a unit conversion:

    $$1 \text{ yr} \times \frac{365 \text{ d}}{\text{yr}} \times \frac{2 \text{ pt}}{\text{d}} \times \frac{1 \text{ qt}}{2 \text{ pt}} \times \frac{1 \text{ gal}}{4 \text{ qt}}, \text{ which is about}$$
    90 gallons per year.

37. Assuming you drive (or ride) about 10 miles per day and drive 350 days each year, the total would be 3500 miles.

39. Answers will vary. Assuming you spend about 10 minutes per day in the shower, you would spend

    $$\frac{10 \text{ min}}{1 \text{ day}} \times \frac{1 \text{ hr}}{60 \text{ min}} \times \frac{365 \text{ days}}{1 \text{ year}} = 60.8 \text{ hr/yr or about}$$
    61 hours per year in the shower.

41. Table 3.1 states that 4 million joules are required for one hour of running, and that an average candy bar supplies 1 million joules of energy. Thus four candy bars would be needed for each hour of running, which means you'd need to eat 24 candy bars to supply the energy for six hours of running.

43. One kilogram of uranium-235 releases 5.6×10^{13} joules, which is 35,000 times as much as the energy released by one kilogram of coal (1.6×10^9 joules): $\dfrac{5.6 \times 10^{13}}{1.6 \times 10^9} = 35,000$.

45. $\dfrac{5 \times 10^7 \text{ J}}{\text{home}} \times \dfrac{1 \text{ liter H}_2\text{O}}{7 \times 10^{13} \text{ J}} = 7 \times 10^{-7} \dfrac{\text{liters}}{\text{home}}$. This is just a little less than one millionth of a liter.

47. $\dfrac{1 \times 10^{20} \text{ J}}{\text{yr}} \times \dfrac{1 \text{ kg uranium}}{5.6 \times 10^{13} \text{ J}} = 1.8 \times 10^6 \dfrac{\text{kg}}{\text{yr}}$. Thus 1.8 million kg of uranium would be needed to supply the annual energy needs of the United States.

49. Since 2 cm represents 100 km, 1 cm represents 50 km. Since there are 5,000,000 cm in 50 km, the scale is 5,000,000 to 1.

51. 1 cm represents 500 km. Since there are 50,000,000 cm in 500 km, the scale is 50,000,000 to 1.

53. In order to scale down each of the actual diameters and distances listed in the table, begin by dividing by 10,000,000,000 (10 billion). Since this produces tiny answers when units of kilometers are used, it is best to convert to a more convenient unit. The table below shows diameters listed in millimeters (rounded to the nearest tenth of a millimeter), and distances listed in meters (rounded to the nearest meter).

Planet	Model Diameter	Model distance from Sun
Mercury	0.5 mm	6 m
Venus	1.2 mm	11 m
Earth	1.3 mm	15 m
Mars	0.7 mm	23 m
Jupiter	14.3 mm	78 m
Saturn	12.0 mm	143 m
Uranus	5.2 mm	287 m
Neptune	4.8 mm	450 m

55. a. The scale on this timeline would be 4.5 billion years to 100 meters, and thus 1 billion years is equivalent to 1/4.5 of 100 meters, which amounts to 22.2 meters.

b. Since 10,000 years is 1/100,000 of one billion years, we only need to divide 22.2 meters by 100,000 to find the distance that goes with 10,000 years. This gives 0.000222 meters, or $0.000222 \text{ m} \times \dfrac{1000 \text{ mm}}{1 \text{ m}} = 0.222 \text{ mm}$.

FURTHER APPLICATIONS

57. $\dfrac{4 \times 10^6 \text{ births}}{\text{yr}} \times \dfrac{1 \text{ yr}}{365 \text{ d}} \times \dfrac{1 \text{ d}}{24 \text{ hr}} \times \dfrac{1 \text{ hr}}{60 \text{ min}} \approx 7.6 \dfrac{\text{births}}{\text{min}}$

59. $\dfrac{32,300 \text{ deaths}}{\text{yr}} \times \dfrac{1 \text{ yr}}{365 \text{ d}} = 88 \dfrac{\text{deaths}}{\text{day}}$

61. $\dfrac{31,700 \text{ fatalities}}{\text{yr}} \times \dfrac{1 \text{ yr}}{365 \text{ d}} \times \dfrac{1 \text{ d}}{24 \text{ hr}} \approx 3.6 \dfrac{\text{fatalities}}{\text{hr}}$

63. There are 315 million Americans, so the amount consumed per person is $\dfrac{3.8 \times 10^{10} \text{ lb}}{3.15 \times 10^8 \text{ people}} = 121$ pounds per person.
This is about 121 lb/12 months = 10 pounds per person per month.

65. a. $1 \text{ cm}^3 \times \left(\dfrac{10^4 \ \mu\text{m}}{1 \text{ cm}}\right)^3 \times \dfrac{1 \text{ cell}}{100 \ \mu\text{m}^3} = 1 \times 10^{10} \text{ cells}$, or 10 billion cells in 1 cm^3.

b. One liter is 1000 milliliters, or 1000 cubic centimeters, so there are 1000 times as many cells in a liter as there are in a cubic centimeter, which comes to 1×10^{13} cells (see part **a**).

c. If one liter of water weighs 1 kilogram, then a person who weighs 70 kg has a volume of 70 liters. From part **b**, we know each liter contains 1×10^{13} cells, and thus we need only multiply that result by 70 to find the number of cells in a human body. This results in 7×10^{14} cells.

67. If we assume an average of three people per household, there are 100 million households in the U.S. (because there are 300 million people). If 4.5 trillion gallons is enough to supply all households for five months, then $\dfrac{4.5 \times 10^{12}}{5} = 9 \times 10^{11}$ gallons is sufficient for one month. Thus each household uses $\dfrac{9 \times 10^{11} \text{ gal}}{1 \times 10^8 \text{ households}} = 9000$ gallons per month. Answers will vary regarding whether this is reasonable, but considering water used for bathing, washing, drinking, cooking, cleaning, and upkeep of the lawn/garden, this is within the ballpark.

69. a. The volume of a sphere is $V = \frac{4}{3}\pi r^3$, and thus the volume of the white dwarf will be $\frac{4}{3}\pi(6400 \text{ km})^3$. Density is mass per unit volume, so the density of the white dwarf will be

$$\frac{2 \times 10^{30} \text{ kg}}{\frac{4}{3}\pi(6400 \text{ km})^3} \times \left(\frac{1 \text{ km}}{10^5 \text{ cm}}\right)^3 = 1821\frac{\text{kg}}{\text{cm}^3}.$$

b. Since a teaspoon is about 4 cubic centimeters, and each cubic centimeter has mass of 1821 kg (part **a**), a teaspoon will have a mass of 7300 kg, which is about the mass of a tank.

c. Following the calculations in part **a**, the density is $\frac{2.8 \times 10^{30} \text{ kg}}{\frac{4}{3}\pi(10 \text{ km})^3} \times \left(\frac{1 \text{ km}}{10^5 \text{ cm}}\right)^3 = 7 \times 10^{11} \frac{\text{kg}}{\text{cm}^3}$, which means one cubic centimeter of this material is more than ten times the mass of Mount Everest.

71. a. $\dfrac{\$1.02 \times 10^{13}/\text{year}}{3.15 \times 10^8 \text{ people}} \approx \$32,400/\text{person/year}$

b. $\dfrac{\$32,381/\text{person}}{1 \text{ year}} \times \dfrac{1 \text{ year}}{365 \text{ day}} \approx \$90/\text{person/day}$

c. $\dfrac{\$6.8 \times 10^{12}}{\$1.02 \times 10^{13}} \approx 0.67 = 67\%$

d. $\dfrac{\$1.7 \times 10^{12}}{\$1.02 \times 10^{13}} \approx 0.17 = 17\%$

e. Total spending: $\dfrac{\$10.2 - \$6.8}{\$6.8} = 0.50 = 50\%$

Health care: $\dfrac{\$1700 - \$918}{\$918} \approx 0.85 = 85\%$

73. Careful measurements show that a penny is about 1.5 mm thick, a nickel 1.8 mm, a dime 1.3 mm, and a quarter 1.7 mm thick. The easiest way to make these measurements is to measure a stack of ten coins, and divide the answer by ten. Although a quarter is almost the thickest coin (among these), its value is 2.5 times as much as the thinnest coin, the dime, and yet it's thickness is not anywhere near 2.5 times as thick as a dime (it's about 1.3 times as thick). So while you'll get more dimes in a stack that is as tall as you are, the value of that stack won't be as much as the value of the stack of quarters – take the quarters.

75. Answers will vary considerably, depending on the amount of light pollution in the area where one might make an estimate.

TECHNOLOGY EXERCISES

83. a. $1 \text{ yr} \times 186,000 \dfrac{\text{mi}}{\text{sec}} \times \dfrac{60 \text{ sec}}{1 \text{ min}} \times \dfrac{60 \text{ min}}{1 \text{ hr}} \times$

$\dfrac{24 \text{ hr}}{1 \text{ day}} \times \dfrac{365 \text{ days}}{1 \text{ yr}} \approx 5.87 \times 10^{12} \text{ miles}$

b. $\dfrac{52 \cdot 51 \cdot 50 \cdot 49 \cdot 48}{5 \cdot 4 \cdot 3 \cdot 2 \cdot 1} = 2,598,960 \text{ hands}$

c. $\dfrac{30,000 \times 10^6 \text{ metric tons}}{6.8 \times 10^9 \text{ persons}} \approx 4.4 \text{ metric tons per person}$

d. $\dfrac{6.0 \times 10^{24} \text{ kg} \times \dfrac{1000 \text{ g}}{1 \text{ kg}}}{1.1 \times 10^{12} \text{ km}^3 \times \left(\dfrac{100,000 \text{ cm}}{1 \text{ km}}\right)^3} \approx 5.5 \dfrac{\text{g}}{\text{cm}}$

e. $14 \times 10^9 \text{ years} \times \dfrac{365 \text{ days}}{1 \text{ yr}} \times \dfrac{24 \text{ hr}}{1 \text{ day}} \times$

$\dfrac{60 \text{ min}}{1 \text{ hr}} \times \dfrac{60 \text{ sec}}{1 \text{ min}} \approx 4.4 \times 10^{17} \text{ sec}$

UNIT 3C

QUICK QUIZ

1. **b.** The best economists can do when making projections into the future is to base their estimates on current trends, but when these trends change, it often proves the predictions wrong.

2. **a.** There are two significant digits in 5.0×10^{-1}, while the other answers have only one.

3. **b.** 1.020 has four significant digits, 1.02 has three, and 0.000020 has only two.

4. **a.** Random errors can either be too high or too low, and averaging three readings is likely to reduce random errors because the high readings typically cancel the low readings.

5. **b.** If you place the thermometers in sunlight, this is a problem with your system of measurement, and the readings will all be too high, which is a systematic error.

6. **c.** Since all the scores are affected the same way (50 points too low), this is an example of a systematic error.

7. **b.** The absolute error in all cases was –50 points, but unless all students had the same score (very unlikely), the relative errors will be different.

8. **a.** Because the scale is able to report your weight to the nearest 1/10 of a pound, it is fairly precise (in comparison to standard bathroom scales), but its accuracy is lacking as it is off by more than 30 pounds (in 146 pounds).

9. **a.** Reporting the debt to the nearest penny is very precise, though the debt is so large that we really don't know it to that level of precision, and thus it is not likely to be very accurate.

10. **a.** Multiplying the gas mileage by the tank capacity will produce the number of miles the car can go on a full tank (290 miles), so the question boils down to this: "How many significant digits should be used when reporting the answer?" Since there are two significant digits in 29 mpg, and three in 10.0 gallons, we should use two significant digits in the answer (use the rounding rule for multiplication), and thus the answer is 290 miles.

DOES IT MAKE SENSE?

7. Does not make sense. Predicting the federal deficit for the next year to the nearest million dollars is nearly impossible.

9. Does not make sense. Unless you have access to an exceptionally precise scale, you could never measure your weight to that level of precision (and even with an excellent scale, your weight changes at the level reported when you sweat, breath, lose a few hairs, etc.).

11. Does not make sense. A typical yard stick may be precise to the nearest 1/16 of an inch, but certainly not to the nearest micrometer.

13. Makes sense. Provided the company's revenue is $1 billion, $1 million would be only 0.1% of revenue, and it wouldn't be considered a large error.

BASIC SKILLS AND CONCEPTS

15. a. 3   b. 88   c. 0
    d. 185   e. 1945   f. 3
    g. 6   h. 1500   i. –14

17. There are four significant digits, and it is precise to the nearest dollar.

19. There is one significant digit, and it is precise to the nearest 10 mph.

21. There are six significant digits, and it is precise to the nearest 0.00001 mile.

23. There are two significant digits, and it is precise to the nearest 1000 seconds.

25. There are seven significant digits, and it is precise to the nearest 0.0001 pound.

27. There are four significant digits, and it is precise to the nearest 0.1 km/s.

29. $45 \times 32.1 = 1440$

31. $231.89 \div 0.034 = 6800$

33. $(2.3 \times 10^5) \times (7.963 \times 10^3) = 1.83149 \times 10^3$

35. Random errors may occur when birds are miscounted, and systematic errors may happen the same bird is are counted more than once (if it leaves and then returns to the region) or a bird that enters the region is not seen.

37. Random errors could occur when taxpayers make honest mistakes or when the income amounts are recorded incorrectly. Systematic errors could occur when dishonest taxpayers report income amounts that are lower than their true income amounts.

39. Random errors could occur when people don't know their actual weight and report an amount that is wrong. Systematic errors could occur when people intentionally lie about their weight by reporting a value that is considerably lower than the true weight.

41. Random errors could occur if the thermometer used to measure the temperatures is read incorrectly, while systematic errors may occur if the thermometer is not calibrated correctly or is in a location that affects its reading, such as direct sunlight.

43. Problem (1) is a random error because the mistakes may be too high or two low. Problem (2) is a systematic error, because the tax payer likely underreported his income.

45. This is a systematic error that results in all altitude readings being 2780 lower than they should be.

47. The absolute error is measured value – true value, or 67.5 in – 68.0 in = –0.5 in. The relative error

   is $\dfrac{\text{measured value} - \text{true value}}{\text{true value}} =$

   $\dfrac{67.5 \text{ in} - 68.0 \text{ in}}{68.0 \text{ in}} = 0.007 = 0.7\%$.

49. The absolute error is 60 mi/hr – 58 mi/hr = 2 mi/hr. The relative error is (60 mi/hr – 58 mi/hr) / (58 mi/hr) = 0.034 = 3.4%.

51. The absolute error is 4.75 oz – 5 oz = –0.25 oz. The relative error is (4.75 oz – 5 oz)/(5 oz) = –0.05 = –5.0%.

53. The absolute error is 24.65 cm – 24.5 cm = 0.15 cm. The relative error is (24.65 cm – 24.5 cm) / (24.5 cm) = 0.006 = 0.6%.

55. The measurement obtained by the tape measure is more accurate because the value is closer to your true height than the value obtained at the doctor's office (1/8 inch versus 0.4 inch). However, the laser at the doctor's office is more precise as it measures to the nearest 0.05 (or 1/20) inch, while the tape measure measures to the nearest 1/8 inch.

57. The digital scale at the gym is both more precise and more accurate. It is more precise because it measures to the nearest 0.01 kg, while the health clinic's scale measures to the nearest 0.5 kg, and it is more accurate because its reported weight is nearer to your true weight than the scale at the health clinic.

59. A calculator gives 12 lb – 0.1 lb = 11.9 lb. This should be rounded to the nearest 1 lb because that's the level of precision of 12 lb. The final answer is 12 lb.

61. A calculator gives 102 mi/0.65 hr = 156.923 mi/hr. However, since 102 has 3 significant digits while 0.65 has only two, we must use only two in the final answer: 160 mi/hr.

63. The least precise measurement is 36 miles, which is precise to the nearest mile, so the answer should be rounded to the nearest mile. 36 mi + 2.2 mi = 38.2 mi, which is 38 mi when rounded to the correct level of precision.

65. The cost is $2.1 million/120,345 people = $17.45 per person. However, as the first measurement has only two significant digits, so must the answer, so we round to $17 per person.

FURTHER APPLICATIONS

67. Random or systematic errors could be present, (for example, a car could be counted twice or not counted). The measurement is not believable with the given precision.

69. Systematic errors could be present if the equipment (such as a transit) used to measure the height is faulty, and random errors could occur in several ways (e.g. if a transit is used, and it wasn't aimed precisely at the top of the building). But even considering these factors, the figure is believable with the given precision.

71. Systematic errors could arise depending on how the area of a lake is defined and how it is measured (should we measure at the current water level, or at the highest recorded water level?), and random errors could easily arise due to changes in the shape of the lake over time, or errors in recording the data. The fact is not believable because (as explained above), it is rather difficult to define how the area of a lake should be measured, let alone measure such a large lake to the nearest square mile.

73. This measurement could have many sources of errors, both random and systematic (for example, people that were killed may not be reported missing and some people that were killed may never be found). The measurement is not believable as the level of precision (to the nearest person) is unattainable, especially in the chaos following a major natural disaster.

75. a. Since each cut can be off by 0.25 inches, the maximum error in either direction will be 20 $\times$ 0.25 in = 5 in. The lengths of the boards can be between 48 in – 5 in = 43 in and 48 in + 5 in = 53 in.

 b. Since each cut can be off by 0.5% or 0.005, the lengths of the boards can be between $48 \times (0.995)^{20} \approx 42.42$ in and $48 \times (1.005)^{20} \approx 53.03$ in.

UNIT 3D

QUICK QUIZ

1. **b.** The 2000 index of 127.9 is $\frac{127.9}{100} = 1.297$ times the 1980 index of 100, so the price of gas in 2000 is 1.297 times as much as the price in 1980.

2. **c.** If you know the price of gas in 1980, you can divide today's price by the 1980 price to get the index number for the current year.

3. **b.** Like most indexes, the CPI is designed to allow one to compare prices from one year to another.

4. **c.** To compare prices in one year to prices in another, one can just divide the CPI from one year by the other – this tells you the factor by which prices in one year differ from prices in another.

5. **c.** The CPI in 2008 was 215.3, and it was 152.4 in 1995. Thus prices in 2008 are 41% more than prices in 1995 (because 215.3/152.4 = 1.41). If the CPI were recalibrated so that 1995 = 100, the new CPI in 2005 would be 141, which is lower than 215.3.

6. **b.** As long as the computers being compared are of comparable power, since the prices have declined, the index would decrease.

7. **a.** If the cost of college has increased faster than the CPI, but one's salary has only kept pace with the CPI, then the cost of college as a percentage of salary has increased, making it more difficult to afford.

8. **a.** If your salary increases faster than the CPI, then it has more purchasing power than in previous years.

9. **c.** Measured in 2006 dollars, gas was less than $1.50 per gallon in 1998–1999, and this is lower than any other time shown on the chart.

10. **c.** In order to reflect the fact that housing prices have tripled from 1985 to 2005, the index would also triple, from 100 to 300.

DOES IT MAKE SENSE?

5. Does not make sense. One must consider the effect of inflation on the price of gas by comparing prices using a price index of, say, 2010 dollars in order to make a valid comparison.

7. Makes sense. A penny in Franklin's day would be worth something on the order of a dollar today.

9. Does not make sense. The reference year that is chosen for an index number does not affect the trends of inflation, so if milk is more expensive in 1995 dollars, it should also be more expensive in 1975 dollars.

BASIC SKILLS AND CONCEPTS

11. To find the index number, take the current price and divide it by the 1980 price: $3.50/$1.22 = 2.869 = 286.9%. The price index number is 286.9 (the percentage sign is dropped).

13. The price index number in 2010 is 232.8, which means prices in 2010 were 232.8% of the prices in 1980. Thus $8 spent on gas in 1980 is equivalent to $8 × 232.8% = $8 × 2.328 = $18.62.

15. To find the fraction of a tank you could purchase, divide the 1970 price by the 2000 price:

$$\frac{36.0¢}{156.0¢} = 0.23 = 23\%,$$ so you could buy 0.23 of a

tank with $10 in 2000.

17. Divide the 2008 CPI by the 1976 CPI to find the factor by which prices in 2008 were larger than prices in 1977: $\frac{CPI_{2008}}{CPI_{1976}} = \frac{215.3}{56.9} = 3.784$. This means prices in 2008 were about 3.784 times larger than prices in 1976, so $20,000 in 1976 is equivalent to $20,000 × $\frac{215.3}{56.9}$ = $75,677 in 2008.

19. Inflation is the relative change in the CPI, which is $\frac{CPI_{2000} - CPI_{1990}}{CPI_{1990}} = \frac{172.2 - 130.7}{130.7} \approx 0.318 =$ 31.8%. Note that this is inflation over the span of ten years (that is, it is not an annual rate of inflation).

21. The price in 2005 dollars is

$$\$0.25 \times \frac{CPI_{2005}}{CPI_{1976}} = \$0.25 \times \frac{195.3}{56.9} = \$0.86$$

23. The price in 1980 dollars is

$$\$9.00 \times \frac{CPI_{1980}}{CPI_{2008}} = \$9.00 \times \frac{82.4}{215.3} = \$3.44$$

25. The purchasing power of a 1976 dollar in terms of 2006 dollars is

$$\$1 \times \frac{CPI_{2006}}{CPI_{1976}} = \$1 \times \frac{201.6}{56.9} = \$3.54.$$

FURTHER APPLICATIONS

27. In Miami, the price would be $300,000 × $\frac{194}{100}$, which is $582,000. In Cheyenne, the price would be $300,000 × $\frac{60}{100}$ = $180,000.

29. In San Francisco, the price would be $250,000 × $\frac{382}{60}$ = $1,591,667. In Boston, the price would be $250,000 × $\frac{358}{60}$ = $1,491,667.

31. Health care spending increased by a percent of

$$\frac{\$2.7 \times 10^{12} - \$85 \times 10^9}{\$85 \times 10^9} = 30.76 = 3076\%,$$

whereas the overall rate of inflation was $\frac{CPI_{2011} - CPI_{1976}}{CPI_{1976}} = \frac{224.9 - 56.9}{56.9} = 2.95 = 295\%.$

33. The relative change in the cost of private colleges was $\dfrac{\$23{,}210 - \$8396}{\$8396} = 1.764 = 176.4\%$. The rate of inflation was $\dfrac{CPI_{2010} - CPI_{1990}}{CPI_{1990}} =$

$\dfrac{218.1 - 130.7}{130.7} = 0.669 = 66.9\%.$

35. As shown in the first line of the table, $0.25 in 1938 dollars is worth $2.78 in 1996 dollars.

37. In 1996, actual dollars are 1996 dollars.

39. The 1979 minimum wage, measured in 1979 dollars, was $2.90. To convert this to 1996 dollars, use the CPI from 1996 and 1979.

$$\$2.90 \times \frac{CPI_{1996}}{CPI_{1979}} = \$2.90 \times \frac{156.9}{72.6} = \$6.27 \,.$$

This agrees with the entry in the table.

41. The purchasing power of the minimum wage was highest in 1968, when its value in 1996 dollars was highest ($7.21). Since all entries in the last column are listed in 1996 dollars, we can make valid comparisons between values in the column.

43. a. An index is usually the ratio of two quantities with the same units (such as prices), which means it has no units.

b. The major league average is $197.36.25, so the index values for each team are:

Team	Index
Boston	$\dfrac{\$339.01}{197.36} \times 100 = 171.8$
NY (Yankees)	$\dfrac{\$338.32}{197.36} \times 100 = 171.4$
St. Louis	$\dfrac{\$223.18}{197.36} \times 100 = 113.1$
Major League Average	$\dfrac{\$197.36}{197.36} \times 100 = 100.0$
Atlanta	$\dfrac{\$169.09}{197.36} \times 100 = 85.7$
San Diego	$\dfrac{\$125.81}{197.36} \times 100 = 63.7$
Arizona	$\dfrac{\$120.96}{197.36} \times 100 = 61.3$

TECHNOLOGY EXERCISES

51. $10 in 2009 had the same buying power as $0.93 in 1920.

53. $1000 in 2008 had the same buying power as $46.91 in 1915.

55. C will be less than D.

UNIT 3E

QUICK QUIZ

1. **a.** 8 refers to the number of patients with severe acne that were cured by the new treatment.

2. **a.** As stated in the text surrounding the table, the old treatment had a better overall treatment rate, even though the new treatment performed better in each category.

3. **b.** Even though Derek had a higher GPA in each of the two years, we can't be sure his overall GPA was higher than Terry's – Simpson's paradox may be at play (see historical note on page 173).

4. **a.** A *false negative* means the test detected no cancer (that's the *negative* part) even though the person has it (the *false* part).

5. **c.** A *false positive* means the test detected steroids (that's the *positive* part), but the person never used steroids (that's the *false* part).

6. **c.** The last row in the column labeled *Tumor is benign* shows the total of those who did not have malignant tumors.

7. **b.** Those who had false negatives (15) and false positives (1485) make up the group who had incorrect test results.

8. **c.** Imagine a test group of 1000 women, where 900 are actually pregnant, and 100 are not. Of those who are pregnant, 1% (that is, 9 women) will test negative. Of those who aren't pregnant, 99% (that is, 99 women) will test negative. Thus 108 women test negative, and 99 of them actually aren't pregnant. Since 99/108 is not 99% (it's closer to 92%), statement **c** is not true (it may be true in other scenarios, but in general, it is not).

9. **c.** Taxpayers in the top 1% would have paid 31% of tax revenue without the tax cuts and paid 37% of tax revenue with the tax cuts, which most likely means they paid more in income taxes.

10. **b**. The figure of $1900 is a taxpayers' savings, not their annual tax payment.

DOES IT MAKE SENSE?

5. Makes sense. Both categories could show an improvement with the new drug even though overall, the old drug may have done a better job.

7. Does not make sense. There are scenarios (see discussion in text concerning mammograms and drug tests, for example) where a positive test may not correlate well with the probability that a bag contains banned materials, even though the test is 98% accurate.

9. Does not make sense. Both sides can make a valid argument for their position (see discussion in text about tax cuts).

BASIC SKILLS AND CONCEPTS

11. a. As shown in the tables, Josh had a higher batting average in the first half of the season.

b. As shown in the tables, Josh had a higher batting average in the second half of the season.

c. Jude had the higher overall batting average (80/200 = .400 versus 85/220 = .386).

d. One person (Josh) performed better in two of two categories (the first and second halves of the season), and yet Jude outperformed Josh over the course of the entire season.

13. a. New Jersey had higher scores in both categories (283 versus 281, and 252 versus 250), but Nebraska had the higher overall average (277 versus 272).

b. It's possible for one state (Nebraska) to score lower in both categories, and yet have a higher overall percentage. This is due to the different racial makeup of the two state's populations. The white students are scoring better than the nonwhite students in both states, and Nebraska has a larger percentage of white students than New Jersey, so they influence the overall scores more heavily than the white students in New Jersey.

c. The verification is an exercise in a weighted average. 87% of the population had an average of 281, and 13% of the population had an average of 250, so the overall average is 0.87(281) + 0.13(250) = 277.

d. As in part **c**, we only have to compute a weighted average: 0.66(283) + 0.34(252) = 272.

e. See part **b**.

15. a. The death rate for whites was 8400/4,675,000 = 0.0018 = 0.18%. For nonwhites, it was 500/92,000 = 0.0054 = 0.54%. Overall, the death rate was 8900/4,767,000 = 0.0019 = 0.19%.

b. The death rate for whites was 130/81,000 = 0.0016 = 0.16%. For nonwhites, it was 160/47,000 = 0.0034 = 0.34%. Overall, the death rate was 290/128,000 = 0.0023 = 0.23%.

c. The death rates for TB in each category (white and nonwhite) were higher in New York City than in Richmond, and yet the overall death rate due to TB was higher in Richmond. The paradox arises because Richmond had a different racial makeup than New York City: it had a much higher percentage of nonwhite residents than New York, and as the death rate for TB was much higher among nonwhites than whites in both cities, the higher percentage of nonwhites in Richmond has a more pronounced influence on the death rate.

17. a. There are 10,000 total participants with tumors in the study (lower right hand entry), and 1% of them (100) have malignant tumors. Since the test is 90% accurate, 90 of those with malignant tumors will be true positives, while the other 10 will be false negatives. 99% of the women (9900) will have benign tumors. 90% of these (8910) will be true negatives, while 10% of these (990) will be false positives. The only numbers in the table not yet verified are the first two entries in the Total column; these are simply the sum of the rows in which they lie.

b. Of the 1080 who have positive mammograms, only 90 are true positives, so only 90/1080 = 8.3% of them actually have cancer.

c. 100 of the women have cancer (malignant tumors), and 90 of those will have a true positive test result, which is 90%. Thus if you really have cancer, the test will detect it 90% of the time.

d. Of the 8920 who have a negative mammogram, only 10 of them are false negatives, which translates to a probability of 10/8920 = 0.11%.

19. a. Beginning with the last row, the sample size is 4000, and 1.5% of them (60) have the disease, while the remaining 3940 do not. The first column shows the 60 with the disease, and the test detects 90% of them (54), but misses 6 of them. The second column shows the 3940 who do not have the disease, and the test says 90% of them (3546) do not have the disease (a negative test), but it says 10% of them (394) do have the disease. The entries in the last column are simply the sum of the rows in which they lie.

b. 54 of the 60 (90%) who have the disease test positive.

19. (continued)

c. 54 of the 448 (12%) who test positive actually have the disease. In part **b**, the answer of 90% is the ratio of those who test positive out of those who have the disease, while the answer of 12% here is the ratio of those who have the disease among those that test positive.

d. As shown in part **c**, if you test positive, you have a 12% chance of actually having the disease. This is 8 times the overall incidence rate (1.5%), which means there is reason for concern, and yet it's still a fairly low chance that you actually have the disease.

FURTHER APPLICATIONS

21. a. Spelman won 10/29 = 34.5% of its home games and 12/16 = 75% of its away games, while Morehouse won 9/28 = 32.1% of its home games and 56/76 = 73.7% of its away games. Since Spelman is better than Morehouse in each category, it's the better team.

b. Morehouse won 65/104 = 62.5% of all the games they played, while Spelman won 22/45 = 48.9% of all the games they played. Based on these figures, Morehouse is the better team.

c. Disregarding the strength of the opponents each college played, it's universally accepted that the team with a higher percentage of wins is the better team, and thus the claim in part **b** makes more sense. Additionally, it's more difficult to win away games than home games, and while both teams have an impressive away-game win percentage, Morehouse compiled its overall record of 62.5% with many more away games.

23. a. Note that 10% of the 5000 at risk people is 500, and this is the sum of the first row in the first table. For the 20,000 people in the general population, note that 0.3% of 20,000 is 60, and this is the sum of the first row in the second table. 95% of the 500 infected people in the at risk group should test positive, and 5% should test negative – take 95% and 5% of 500 to verify the first row entries in the first table. 95% of the 4500 at risk people who aren't infected should test negative, while 5% should test positive – take 95% and 5% of 4500 to verify the second row in the first table.

A similar process will show that the numbers in the second table are also correct.

b. 475/500 = 95% of those with HIV test positive. 475/700 = 67.9% of those who test positive have HIV. The percentages are different because the number of people in each group (those who have HIV and those who test positive) is different.

c. As shown in part **b**, if you test positive, and you are in the at risk group, you have a 67.9% chance of carrying the disease. This is much greater than the overall percentage of those who have the disease (10%), which means if you test positive, you should be concerned.

d. 57/60 = 95% of those with HIV test positive. 57/1054 = 5.4% of those who test positive have HIV. The two percentages are different because the number of people in each group (those who have HIV and those who test positive) is different.

e. As shown in part **d**, if you test positive, and you are in the general population, you have a 5.4% chance of carrying the disease. This is larger than the overall percentage of those who have the disease (0.3%), but it's still small enough that it's best to wait for more tests before becoming alarmed.

25. The data suggest a hiring preference for women because a higher percentage of women were hired in each of the two categories (white-collar and blue-collar jobs). On the other hand, if one looks at the overall picture, it turns out a higher percentage of male applicants was hired. (There were 20% of 200 applicants = 40 female workers hired for white-collar jobs, and 85% of 100 applicants = 85 female workers hired for blue-collar jobs. Thus there were 125 workers hired from a pool of 300 applicants, which is 125/300 = 41.7%. There were 15% of 200 applicants = 30 male workers hired for white-collar jobs, and 75% of 400 applicants = 300 male workers hired for blue-collar jobs. Thus there were 330 workers hired from a pool of 600 applicants, which is 330/600 = 55%.) Females were hired at a greater rate in both categories of workers, and yet males were hired at a greater rate overall. The paradox can be resolved by realizing that there were many more males hired for the blue-collar jobs, which are more abundant than the white-collar jobs, and thus the male-dominated blue-collar jobs have a greater influence on the overall hiring rate.

27. a.

	Republicans	Democrats	Totals
Women	4	16	20
Men	41	37	78
Totals	45	53	98

b. $\dfrac{4}{98} = 0.041 = 4.1\%$

c. $\dfrac{4}{20} = 0.20 = 20\%$

27. (continued)

d. $\frac{4}{45} = 0.089 = 8.9\%$

e. In the part **b** we computed the percentage from 20 women; in part **c**, we computed the percentage from 45 Republicans.

f. The percentage of women among all senators is $\frac{20}{100} = 0.20 = 20\%$, while the percentage of women among Democrats is $\frac{16}{53} = 0.302 = 30.2\%$.

29. a. The absolute difference in savings is $12,838 – $211 = $12,627. A single person earning $41,000 will save $211/$41,000 = 0.5% of his income. A single person earning $530,000 will save $12,838/$530,000 = 2.4% of his income.

b. The absolute difference in savings is $13,442 – $1208 = $12,234. A married couple earning $41,000 will save $1208/$41,000 = 2.9% of their income. A married couple earning $530,000 will save $13,442/$530,000 = 2.5% of their income.

UNIT 4A

QUICK QUIZ

1. **a**. Evaluating your budget allows you to look critically at your cash flow, which affects personal spending.

2. **a**. Your cash flow is determined by the amount of money you earn (income), and the amount you spend.

3. **b**. If your cash flow is negative, you are spending more than you take in.

4. **c**. You should prorate all once-per-year expenses and include them in your monthly budget.

5. **b**. Housing typically costs about one-third of your income.

6. **c**. As a percentage of income, health care expenses are not too alarming, though they grow rapidly as one ages.

7. **a**. You can't save without money left over at the end of the month, which corresponds to a positive cash flow.

8. **c**. Trey spends about $8 per day on cigarettes, which comes close to $250 per month.

9. **b**. Each week, you use about 22 gallons of gas, because (400 mi)/(18 mi/gal) ≈ 22 gal. With gas at $4/gal, this comes to $88/week, or $4576/year. A car that gets 50 mi/gal will use about one-third of the gas used by your old car (because 50 is almost three times as large as 18), which means it will cost about $1500/year for gas. The difference between these two values is the amount you save, around $3000.

10. **b**. You earn about $12,000 more with a bachelor's degree, which is just over $350,000 in earnings in 30 years.

DOES IT MAKE SENSE?

7. Does not make sense. All the smaller expenses do add up, and they have a significant influence on your budget.

9. Makes sense. When prorated as a per-month expense, an $1800 vacation costs $150 per month.

11. Does not make sense. Divide $15,000 by 365 (days per year) to find that an annual expense of $15,000 comes to $41 per day. Pizza and a soda do not cost that much.

BASIC SKILLS AND CONCEPTS

13. Maria spends $20 each week on coffee, and using 52 weeks per year, this comes to $1040 per year. She spends $130 each month on food, which is $1560 per year. The amount she spends on coffee is $1040/$1560 = 67% of the amount she spends on food.

15. Suzanne spends $85 per month for her cell phone usage, which is $1020 per year. This is $1020/$200 = 510% of the amount she spends on health insurance.

17. Sheryl spends $9 each week on cigarettes, which comes to $468 per year. She spends $30 each month on dry cleaning, or $360 per year. The amount she spends on cigarettes is $468/$360 = 130% of the amount she spends on dry cleaning.

19. Vern spends $21 per week on beer, which amounts to $1092 per year. This is $1092/$700 = 156% of the amount he spends on textbooks.

21. Since 18%/12 = 1.5%, you spend 1.5% of $650 on interest each month, which is 0.015 × $650 = $9.75. Over the span of a year, you'll spend $117 on interest.

23. Vic's balance is $2200 − $300 = $1900. He spends 3% of $1900, or 0.03 × $1900 = $57 each month on interest. This comes to $684 per year.

25. Sara spends $6000 per semester, or $12,000 per year, which is $12,000/12 = $1000 per month.

27. Lan spends $650 × 2 = $1300 per year on automobile insurance, $125 × 12 = $1500 per year for health insurance, and $400 per year on life insurance. The total cost is $3200, which means he spends $3200/12 = $266.67 per month.

29. Raul spends $200 + $245 + $100 + $300 = $845 each year on these contributions, or $845/12 = $70.42 per month.

31. Income = $600 × 12 + $400 × 12 + $5000 = $17,000 per year. Expenses = $450 × 12 + $50 × 48 + $3000 × 2 + $100 × 48 = $18,600 per year. Thus annual cash flow is $17,000 − $18,600 = −$1600, and monthly cash flow is −$1600/12 = −$133.

33. Income = $2300 × 12 = $27,600. Expenses = $800 × 12 + $90 × 48 + $125 × 12 + $360 × 2 + $400 × 2 + $25 × 48 + $400 × 12 + $85 × 12 = $23,960. Thus annual cash flow is $27,600 − $23,960 = $3640, and monthly cash flow is $3640/12 = $303.

35. The woman spends $900/$3200 = 28% of her income on rent (i.e. housing), which is below average.

37. The man spends $200/$3600 = 5.6% on health care, which is about average.

39. The couple spends $600/$4200 = 14%$ on health care, which is above average.

41. The cost of using the old car is computed below.

    Gas:

    $$\frac{250 \text{ mi}}{\text{wk}} \times \frac{1 \text{ gal}}{21 \text{ mi}} \times \frac{\$3.50}{1 \text{ gal}} \times \frac{52 \text{ wk}}{\text{yr}} \times 5 \text{ yr} = \$10,833.$$

    Insurance: $\frac{\$400}{\text{yr}} \times 5 \text{ yr} = \$2000.$

    Repairs: $\$75,000 = Pe^{0.045 \times 35}.$

    Total: $\$10,833 + \$2000 + \$7500 = \$20,333.$

    The cost for using the new car is computed below.

    Gas:

    $$\frac{250 \text{ mi}}{\text{wk}} \times \frac{1 \text{ gal}}{45 \text{ mi}} \times \frac{\$3.50}{1 \text{ gal}} \times \frac{52 \text{ wk}}{\text{yr}} \times 5 \text{ yr} = \$5056.$$

    Insurance: $\frac{\$800}{\text{yr}} \times 5 \text{ yr} = \$4000.$

    Purchase Price: $16,000.

    Total: $\$5056 + \$4000 + \$16,000 = \$25,056.$

    Using the old car is less expensive, but not by a large factor. After five years, if you use the old car, you'll have a junky car, whereas if you bought the new car, you'd have a nice car after five years.

43. If you buy the car for $22,000, and sell it three years later for $10,000, you will have spent $12,000. If you lease the car for three years (36 months) you will spend $1000 + $250 \times 36 = $10,000$. It is cheaper to lease in this case.

45. At the in-state college, you will spend $4000 + $700 \times 12 = $12,400$ each year. At the out-of-state college, you will spend $6500 + $450 \times 12 = $11,900$. It will cost less out-of-state.

47. He will earn $(\$66,196 - \$40,447) \times 40 = \$1,029,960$ more.

49. A man earns approximately $\frac{\$66,196 - \$49,108}{\$49,108} = 35\%$ more than a woman in a year. Over 40 years, a man will earn $(\$66,196 - \$49,108) \times 40 = \$683,520$ more.

51. If you take the course, you will spend $1500, and use 150 hours of time that could be used to earn money in a job. If you work those 150 hours, you will earn $1500. Therefore, the net cost will be $1500 + $1500 = $3000.

FURTHER APPLICATIONS

53. Note that it costs $10 less per month to operate the new dryer, and thus in 62 months, you would have saved $620, which is long enough to pay for the new dryer.

55. a. If you didn't have the policy, you'd pay all of the costs of the claims, which is $450 + $925 = $1375. If you did have the policy, you'd pay two years of premiums $(2 \times \$550 = \$1100)$, all of the $450 claim, and $500 of the $925 claim, for a grand total of $2050.

    b. Without the insurance policy, your cost would be $450 + $1200 = $1650. With the policy, you'd pay premiums for two years, and $200 on each claim for a total of $2 \times \$650 + 2 \times \$200 = \$1700$.

    c. Without the policy, your cost would be $200 + $1500 = $1700. With the policy, you'd pay premiums for two years, $200 for the first claim, and $1000 for the second claim, for a total of $2 \times \$300 + \$200 + \$1000 = \$1800$.

57. a. You'll pay $25 for each of the office visits, $200 for each trip to the emergency room, and $1000 for the surgery, plus one year of premiums, for a grand total of $\$25 \times 3 + \$200 \times 2 + \$1000 + \$350 \times 12 = \$5675$.

    b. If you did not have the insurance policy, your cost is the total of the numbers in the second column, which is $8220.

59. – 61. Answers will vary.

UNIT 4B

QUICK QUIZ

1. **b.** Compound interest always yields a greater balance than simple interest when the APR is the same.

2. **a.** If you begin with a principle P, and add 5% to its value after one year, the result is $P + 0.05P = 1.05P$, which means the principle increases by a factor of 1.05 each year.

3. **c.** The compound interest formula states $A = P \times (1 + \text{APR})^Y$. With an APR of 6.6%, the balance after five years would be $A = P \times (1 + 0.066)^5$, which means the account increases in value by a factor of 1.066^5.

4. **a.** The APR of 4% is divided evenly into four parts so that each quarter (three months), the account increases in value by 4%/4 = 1%.

5. **c.** Compounding interest more often always results in a greater annual percentage yield.

6. **b.** The APY is the same as the APR with annual compounding, but it's larger in all other cases.

7. **a.** After 20 years, the account earning 4% APR will have grown by a factor of $1.04^{20} = 2.19$, whereas the account earning 2% APR will have grown by a factor of $1.02^{20} = 1.49$. Thus the 4% APR account will have earned about one and a half times as much interest.

8. **a.** The continuously compounded interest formula reads $A = P \times e^{(APR \times Y)}$, and thus after two years, the balance is $A = \$500 \times e^{(0.06 \times 2)} = \$500 \times e^{0.12}$.

9. **a.** The compound interest formula assumes a constant APR for as many years as you have your money invested in an account.

10. **c.** Bank accounts that earn compound interest grow by larger amounts as time goes by, and this is a hallmark of exponential growth.

DOES IT MAKE SENSE?

9. Does not make sense. You earn more interest under compound interest when the APR is the same.

11. Does not make sense. A bank with an APR of 5.9% compounded daily is a better deal than a bank with 6% APR compounded annually (all other things being equal) because the APY in the first case is greater.

13. Makes sense. One's bank account grows based on the APY, and an APR of 5% could certainly result in an APY of 5.1% if the number of compounding periods per year was just right.

BASIC SKILLS AND CONCEPTS

15. $2^3 = 2 \times 2 \times 2 = 8$

17. $4^3 = 64$

19. $16^{1/2} = \sqrt{16} = 4$

21. $64^{-1/3} = \dfrac{1}{64^{1/3}} = \dfrac{1}{\sqrt[3]{64}} = \dfrac{1}{4}$

23. $3^4 \div 3^2 = 3^{4-2} = 3^2 = 9$

25. $25^{1/2} \div 25^{-1/2} = 25^{1/2-(-1/2)} = 25^1 = 25$

27. $x = 12$ (Add 3 to both sides.)

29. $z = 16$ (Add 10 to both sides.)

31. $p = 4$ (Divide both sides by 3.)

33. $z = 4$ (Add 1, and divide both sides by 5.)

35. $3x - 4 = 2x + 6 \Rightarrow 3x = 2x + 10 \Rightarrow x = 10$

37. $3a + 4 = 6 + 4a \Rightarrow 3a - 2 = 4a \Rightarrow a = -2$

39. $6q - 20 = 60 + 4q \Rightarrow 6q = 80 + 4q \Rightarrow 2q = 80 \Rightarrow q = 40$

41. $t/4 + 5 = 25 \Rightarrow t/4 = 20 \Rightarrow t = 80$

43. $x^2 = 25 \Rightarrow (x^2)^{1/2} = 25^{1/2} \Rightarrow x = \sqrt{25} = 5$ Another solution is $x = -5$.

45. $(x-4)^2 = 36 \Rightarrow \left((x-4)^2\right)^{1/2} = 36^{1/2} \Rightarrow$ $x - 4 = \sqrt{36} \Rightarrow x = 4 + 6 = 10$ Another solution is $x = -2$.

47. $(t/3)^2 = 16 \Rightarrow \left((t/3)^2\right)^{1/2} = 16^{1/2} \Rightarrow$ $t/3 = \sqrt{16} \Rightarrow t = 3 \times 4 = 12$ Another solution is $t = -12$.

49. $u^9 = 512 \Rightarrow (u^9)^{1/9} = 512^{1/9} \Rightarrow u = \sqrt[9]{512} = 2$

51. You'll earn 4% of $700, or $0.04 \times \$700 = \$28$, each year. After five years, you will have earned $5 \times \$28 = \$140$ in interest, so that your balance will be $840.

53. Each year, you'll earn 3.5% of $3200 = $0.035 \times \$3200 = \$112$. After five years, you will have earned $5 \times \$112 = \$560$ in interest, and your balance will be $3760.

55.

Year	Suzanne Interest	Suzanne Balance	Derek Interest	Derek Balance
0	-----	$3000	-----	$3000
1	$75	$3075	$75	$3075
2	$75	$3150	$77	$3152
3	$75	$3225	$79	$3231
4	$75	$3300	$81	$3311
5	$75	$3375	$83	$3394

Suzanne's balance increases by $375, or 12.5% of its original value. Derek's increases by $394, or 13.1% of its original value. (Note on values shown in the table: the balance in one year added to next year's interest does not necessarily produce next year's balance due to rounding errors. Each value shown is rounded correctly to the nearest cent).

57. $A = \$10,000(1 + 0.04)^{10} = \$14,802.44$

59. $A = \$15,000(1 + 0.032)^{25} = \$32,967.32$

61. $A = \$5000(1 + 0.031)^{12} = \$7212.30$

63. $A = \$10,000\left(1+\dfrac{0.02}{4}\right)^{4\cdot10} = \$12,207.94$

65. $A = \$25,000\left(1+\dfrac{0.03}{365}\right)^{365\cdot5} = \$29,045.68$

67. $A = \$2000\left(1+\dfrac{0.05}{12}\right)^{12\cdot15} = \$4227.41$

69. $A = \$25,000\left(1+\dfrac{0.037}{4}\right)^{4\cdot30} = \$75,472.89$

71. The APY is the relative increase in the balance of a bank account over the span of one year. The easiest way to compute it when the APR is known is to find the one-year balance, compute the relative increase, and express the answer as a percent. When the principal is not given, any amount can be used – here, we will use \$100.

   1-year balance: $A = \$100\left(1+\dfrac{0.031}{365}\right)^{365\cdot1} =$ \$103.15. The relative increase is \$3.15/\$100 = 3.15%, so the APY is 3.15%.

73. See Exercise 71 for details. The one-year balance is $A = \$100\left(1+\dfrac{0.0123}{12}\right)^{12\cdot1} = \$101.24$. The relative increase (APY) is \$1.24/\$100 = 1.24%.

75. The one, five, and twenty-year balances are:

   One year: $A = \$10,000e^{0.035\times1} = \$10,356.20$.

   Five years: $A = \$10,000e^{0.035\times5} = \$11,912.46$.

   Twenty years: $A = \$10,000e^{0.035\times20} = \$20,137.53$.

   To compute the APY, find the relative increase in the balance of the account after one year.
   APY = \$356.20/\$10,000 = 3.56%.

77. The one, five, and twenty-year balances are:

   One year: $A = \$7000e^{0.045\times1} = \$7322.20$.

   Five years: $A = \$7000e^{0.045\times5} = \$8766.26$.

   Twenty years: $A = \$7000e^{0.045\times20} = \$17,217.22$.

   To compute the APY, find the relative increase in the balance of the account after one year.
   APY = \$322.20/\$7000 = 4.60%.

79. The one, five, and twenty-year balances are:

   One year: $A = \$3000e^{0.06\times1} = \$3185.51$.

   Five years: $A = \$3000e^{0.06\times5} = \$4049.58$.

   Twenty years: $A = \$3000e^{0.06\times20} = \$9960.35$.

   To compute the APY, find the relative increase in the balance of the account after one year.
   APY = \$185.51/\$3000 = 6.18%.

81. Solve $\$25,000 = P(1+0.05)^8$ for P to find
   $P = \dfrac{\$25,000}{(1+0.05)^8} = \$16,920.98$.

83. Solve $\$25,000 = P\left(1+\dfrac{0.06}{12}\right)^{12\cdot8}$ for P to find
   $P = \dfrac{\$25,000}{\left(1+\dfrac{0.06}{12}\right)^{12\cdot8}} = \$15,488.10$.

85. Solve $\$120,000 = P\left(1+\dfrac{0.04}{365}\right)^{365\cdot15}$ for P to find
   $P = \dfrac{\$120,000}{\left(1+\dfrac{0.04}{365}\right)^{365\cdot15}} = \$65,859.56$.

87. Solve $\$120,000 = P\left(1+\dfrac{0.028}{4}\right)^{4\cdot15}$ for P to find
   $P = \dfrac{\$120,000}{\left(1+\dfrac{0.028}{4}\right)^{4\cdot15}} = \$78,961.07$.

FURTHER APPLICATIONS

89. After 10 years, Chang has \$705.30; after 30 years, he has \$1403.40. After 10 years, Kio has \$722.52; after 30 years, she has \$1508.74. Kio has \$17.22, or 2.4% more than Chang after 10 years. She has \$105.34, or 7.5% more than Chang after 30 years.

91. To compute the APY, find the one-year balance in the account, using any principal desired (\$100 is used here), and then compute the relative increase in the value of the account over one year.

   One-year balance, compounded quarterly:
   $$A = \$100\left(1+\dfrac{0.066}{4}\right)^{4\cdot1} = \$106.77.$$
   APY = \$6.77/\$100 = 6.77%.

   One-year balance, compounded monthly:

91. (continued)

$$A = \$100\left(1+\frac{0.066}{12}\right)^{12\cdot1} = \$106.80.$$

APY = $6.80/$100 = 6.80%.

One-year balance, compounded daily:

$$A = \$100\left(1+\frac{0.066}{365}\right)^{365\cdot1} = \$106.82.$$

APY = $6.82/$100 = 6.82%.

As the number of compounding periods per year increases, so does the APY, though the rate at which it increases slows down.

93.

Year	Account 1 Interest	Account 1 Balance	Account 2 Interest	Account 2 Balance
0	-----	$1000	-----	$1000
1	$55	$1055	$57	$1057
2	$58	$1113	$59	$1116
3	$61	$1174	$63	$1179
4	$65	$1239	$67	$1246
5	$68	$1307	$70	$1317
6	$72	$1379	$74	$1391
7	$76	$1455	$79	$1470
8	$80	$1535	$83	$1553
9	$84	$1619	$87	$1640
10	$89	$1708	$93	$1733

Account 1 has increased in value by $708, or 70.8%. Account 2 has increased by $733, or 73.3%. (Note on values shown in the table: the balance in one year added to next year's interest does not necessarily produce next year's balance due to rounding errors. Each value shown is rounded correctly to the nearest dollar).

95. a. After 5 years, Rosa has $3000(1.04)^5 = $3649.96; after 20 years, she has $3000(1.04)^{20} = $6573.37. Julian has $2500(1.05)^5 = $3190.70 after 5 years, and he has $2500(1.05)^{20} = $6633.24 after 20 years.

b. For Rosa, ($3649.96 − $3000.00)/$3649.96 = 18% of the balance after 5 years is interest and ($6573.37 − $3000.00)/$6573.37 = 54% of the balance after 20 years is interest. For Julian, ($3190.70 − $2500.00)/$3190.70 = 22% of the balance after 5 years is interest and ($6633.24 − $2500.00)/$6633.24 = 62% of the balance after 20 years is interest.

c. The longer the investment time, the higher the overall returns.

97. For Plan A, solve $120,000 = P(1.05)^{30}$ for P to find $P = \dfrac{\$120,000}{(1.05)^{30}} = \$27,765.29$.

For Plan B, solve $120,000 = Pe^{0.048\times30}$ for P to find $P = \dfrac{\$120,000}{e^{0.048\times30}} = \$28,431.33$.

99. To obtain an exact solution for this problem, one must use logarithms (studied in a later chapter). If your initial investment P is to triple, we must solve $3P = P(1.08)^t$ for t in order to find out how long it takes. Begin by dividing both sides by P (it will cancel from both sides), and then take the logarithm of both sides: $\log 3 = \log(1.08)^t$. A property of logarithms says the exponent t can be moved in front of the logarithm, as such: log 3 = t log(1.08). Now divide both sides by log 1.08 to arrive at the answer: t = (log 3)/(log 1.08) = 14.3 years. Thus it will take about 14.3 years for your investment to triple in value. Another way to come to the solution is to try various values of t in $(1.08)^t$ – you will have found a solution when $(1.08)^t$ is about 3.

101. See Exercise 99. Solve $100,000 = $1000(1.07)^t$ for t. Begin by dividing both sides by $1000; this yields $100 = (1.07)^t$. Now take the logarithm of both sides, and apply a property of logarithms to get log 100 = t log 1.07. Divide both sides by log 1.07 to find t = (log 100)/(log 1.07) = 68.1 years. Trial and error may also produce the same result if you are patient.

103. a. Yes, at the end of the year, the account balance will be $A = \$50,000\left(1+\dfrac{0.055}{12}\right)^{12\cdot1} = \$52,820.39$, and the interest of $2820.39 is sufficient for the scholarship.

b. No, at the end of the year, the account balance will be $A = \$50,000\left(1+\dfrac{0.048}{12}\right)^{12\cdot1} = \$52,453.51$, and the interest of $2453.51 is not sufficient for the scholarship.

c. About 4.9%.

TECHNOLOGY EXERCISES

109. a. FV(0.10,5,0,100) = $161.05

b. FV(0.02,535,0,224) = $8,940,049.24

111. a. FV(0.03/12,60,0,5000) = $5808.08

b. FV(0.045/12,30*12,0,800) = $3078.16

c. FV(0.0375/365,365*50,0,1000) = $6520.19

113. a. $e^{3.2} = 24.5325$

b. $e^{0.065} = 1.0672$

c. $100 \times e^{0.04} = 104.0811$. So the APY is $4.081/$100 = 4.0811%.

UNIT 4C

QUICK QUIZ

1. **a.** As the number of compounding periods n increases, so does the APY, and a higher APY means a higher accumulated balance.

2. **c.** As long as the savings plan in which you are investing money has a positive rate of return, the balance will increase as time increases.

3. **c.** The total return on a five-year investment is simply the percent change of the balance over five years.

4. **b.** The annual return is the APY that gives the same overall growth in five years as the total return.

5. **a.** If you deposit $100 per month for ten years (120 months), you have deposited a total of $12,000 into the savings plan. Since the balance was $22,200, you earned $22,200 – $12,000 = $10,200 in interest.

6. **a.** If you can find a low risk, high return investment, that's the best of both worlds, and adding high liquidity (where it's easy to access your money) is like icing on the cake. Good luck.

7. **b.** Market capitalization is the share price of a company times the number of outstanding shares. Company B has the greatest market capitalization of $9 per share × 10,000,000 shares = $90,000,000, which is greater than the market capitalization of $10,000,000 for companies A and C.

8. **a.** The P/E ratio is the share price divided by earnings per share over the past year.

9. **b.** If the bond is selling at 103 points, it is selling at 103% of its face value, which is $1.03 \times \$5000 = \$5150$.

10. **c.** The one-year return is simply a measure of how well the mutual fund has done over the last year, and it could be higher or lower than the three-year return (which measures how well the fund has done over the last three years).

DOES IT MAKE SENSE?

9. Does not make sense. With an APR of 4%, the savings plan formula gives a balance of

$$A = \frac{\$25\left[\left(1+\dfrac{0.04}{12}\right)^{12\cdot30}-1\right]}{\left(\dfrac{0.04}{12}\right)} = \$17,351 \text{ after } 30$$

years, which probably wouldn't cover expenses for even one year of retirement (and certainly interest earned on that balance would not be enough for retirement). You should plan to save considerably more than $25 per month for a retirement plan.

11. Does not make sense. Stocks can be a risky investment (you might lose all of your money), and most financial advisors will tell you to diversify your investments.

13. Does not make sense. Stocks are considered a risky investment (especially in the short term). In general, an investment with a greater return is even more risky, and thus the claim that there is no risk at all is dubious, at best.

BASIC SKILLS AND CONCEPTS

15. $$A = \frac{\$150\left[\left(1+\dfrac{0.03}{12}\right)^{12}-1\right]}{\left(\dfrac{0.03}{12}\right)} = \$1824.96$$

17. $$A = \frac{\$400\left[\left(1+\dfrac{0.04}{12}\right)^{12\times3}-1\right]}{\left(\dfrac{0.04}{12}\right)} = \$15,272.62$$

19. You save for 40 years, so the value of the IRA is

$$A = \frac{\$75\left[\left(1+\frac{0.05}{12}\right)^{12\times40}-1\right]}{\left(\frac{0.05}{12}\right)} = \$114,451.51$$

Since you deposit $75 each month for 40 years, your total deposits are $\frac{\$75}{mo}\times\frac{12\ mo}{yr}\times40\ yr = \$36,000$. The value of the account is just over three times the amount you deposited.

21. $$A = \frac{\$300\left[\left(1+\frac{0.035}{12}\right)^{12\times18}-1\right]}{\left(\frac{0.035}{12}\right)} = \$90,091.51$$

You have deposited $\frac{\$300}{mo}\times\frac{12\ mo}{yr}\times18\ yr = \$64,800$, which is a little less than three-fourths of the value of the account.

23. Solve $$\$85,000 = \frac{PMT\left[\left(1+\frac{0.05}{12}\right)^{12\times15}-1\right]}{\left(\frac{0.05}{12}\right)}$$ for

PMT. $$PMT = \frac{\$85,000\left(\frac{0.05}{12}\right)}{\left[\left(1+\frac{0.05}{12}\right)^{12\times15}-1\right]} = \$318.01.$$

You should deposit $318.01 each month.

25. Solve $$\$15,000 = \frac{PMT\left[\left(1+\frac{0.055}{12}\right)^{12\times3}-1\right]}{\left(\frac{0.055}{12}\right)}$$ for

PMT. $$PMT = \frac{\$15,000\left(\frac{0.055}{12}\right)}{\left[\left(1+\frac{0.055}{12}\right)^{12\times3}-1\right]} = \$384.19.$$

You should deposit $384.19 each month.

27. First, find out how large your retirement account needs to be in order to produce $100,000 in interest (at 6% APR) each year. You want 6% of the total balance to be $100,000, which means the total balance should be $100,000/0.06 = $1,666,667. Now solve

$$\$1,666,667 = \frac{PMT\left[\left(1+\frac{0.06}{12}\right)^{12\times30}-1\right]}{\left(\frac{0.06}{12}\right)}$$ for

PMT.

$$PMT = \frac{\$1,666,667\left(\frac{0.06}{12}\right)}{\left[\left(1+\frac{0.06}{12}\right)^{12\times30}-1\right]} = \$1659.18.$$ You should deposit $1659.18 each month.

29. The total return is the relative change in the investment, and since you invested $6000 (100 shares at $60 per share), the total return is $\frac{\$9400-\$6000}{\$6000} = 0.567 = 56.7\%$. The annual return is the APY that would give the same overall growth in five years, and the formula for computing it is $\left(\frac{A}{P}\right)^{(1/Y)}-1$. Thus the annual return is $\left(\frac{\$9400}{\$6000}\right)^{(1/5)}-1 = 0.094 = 9.4\%$.

31. The total return is $\frac{\$11,300-\$6500}{\$6500} = 73.8\%$. The annual return is $\left(\frac{\$11,300}{\$6500}\right)^{(1/20)}-1 = 2.8\%$.

33. The total return is $\frac{\$2000-\$3500}{\$3500} = -42.9\%$. The annual return is $\left(\frac{\$2000}{\$3500}\right)^{(1/3)}-1 = -17.0\%$.

35. The total return is $\frac{\$12,600-\$7500}{\$7500} = 68.0\%$. The annual return is $\left(\frac{\$12,600}{\$7500}\right)^{(1/10)}-1 = 5.3\%$.

37. $300 invested in stocks would be worth $\$300(1+0.063)^{112} = \$281,091$. For bonds and cash, the investments would be worth $\$300(1+0.02)^{112} = \$2756$, and $\$300(1+0.009)^{112} = \$818$, respectively.

39. a. The symbol for Intel stock is INTC.

 b. Intel stock closed at $15.48 per share yesterday.

 c. The total value of shares traded today is $15.48/share × 64 million shares = $990 million.

 d. 64 million/5580 million = 1.1% of all Intel shares have been traded so far today.

 e. You should expect a dividend of 100 × $15.48 × 0.0362 = $56.04.

 f. The earnings per share is $15.48/19.77 = $0.783 = $0.78/share.

 g. Intel earned a total profit of $0.783/share × 5580 million shares = $4369 million, so Intel earned a total profit of about $4.37 billion.

41. a. The earnings per share for COSTCO is $48.30/17.25 = $2.80/share.

 b. The stock is slightly overpriced.

43. a. The earnings per share for IBM is $101.89/11.30 = $9.02/share.

 b. The stock is priced just about right.

45. Answers will vary.

47. The current yield on a bond is its annual interest payment divided by its current price. A $1000 bond with a coupon rate of 2% pays $1000 × 0.02 = $20 per year in interest, so this bond has a current yield of $20/$950 = 2.11%.

49. The annual interest payment on this bond is $1000 × 0.055 = $55, so its current yield is $55/$1100 = 5%.

51. The price of this bond is 105% × $1000 = $1050, and it has a current yield of 3.9%, so the annual interest earned is $1050 × 0.039 = $40.95.

53. The price of this bond is 114.3% × $1000 = $1143, and its current yield is 6.2%, which means the annual interest will be $1143 × 0.062 = $70.87.

55. a. You would buy $5000/$10.93 = 457.46 shares.

 b. Your investment would be worth $5000 × (1 + 0.0320)^5 = $5852.86.

 c. Your investment would be worth $5000 × (1 + 0.0369)^{10} = $7183.54.

FURTHER APPLICATIONS

57. After 10 years, the balance in Yolanda's account is

$$A = \frac{\$200\left[\left(1+\frac{0.05}{12}\right)^{12\times10} - 1\right]}{\left(\frac{0.05}{12}\right)} = \$31,056.46.$$ She

deposits $200 per month, so her total deposits are $\frac{\$200}{mo} \times \frac{12\ mo}{yr} \times 10\ yr = \$24,000$. Zach's account is worth

$$A = \frac{\$2400\left[\left(1+\frac{0.05}{1}\right)^{1\times10} - 1\right]}{\left(\frac{0.05}{1}\right)} = \$30,186.94.$$ He

deposits $2400 per year, so his total deposits are $\frac{\$2400}{yr} \times 10\ yr = \$24,000$. Yolanda comes out ahead even though both deposited the same amount of money because the interest in her account is compounded monthly, while Zach's is compounded yearly, which means Yolanda enjoys a higher APY.

59. After 10 years, the balance in Juan's account is

$$A = \frac{\$400\left[\left(1+\frac{0.06}{12}\right)^{12\times10} - 1\right]}{\left(\frac{0.06}{12}\right)} = \$65,551.74.$$ His

total deposits are $\frac{\$400}{mo} \times \frac{12\ mo}{yr} \times 10\ yr = \$48,000$. The balance in Maria's account is

$$A = \frac{\$5000\left[(1+0.065)^{10} - 1\right]}{(0.065)} = \$67,472.11.$$ Her

total deposits are $\frac{\$5000}{yr} \times 10\ yr = \$50,000$. Maria comes out ahead because she has a higher APR (and APY, it turns out), and she deposits more money over the course of the 10 years.

61. Your balance will be $A = \dfrac{\$50\left[\left(1+\frac{0.07}{12}\right)^{12\cdot15} - 1\right]}{\left(\frac{0.07}{12}\right)}$

= $15,848.11, so you won't reach your goal.

63. Your balance will be

$$A = \frac{\$100\left[\left(1+\frac{0.06}{12}\right)^{12\cdot15} - 1\right]}{\left(\frac{0.06}{12}\right)} = \$29,081.87,$$ so

you won't reach your goal.

65. The total return is $\dfrac{\$8.25 - \$6.05}{\$6.05} = 36.4\%$ per share. Incidentally, this is also the annual return (because you bought the stock a year ago), and it doesn't matter how many shares you bought.

67. The total return was $\dfrac{\$22,000,000 - \$5000}{\$5000} =$ 439,900%. The annual return was $\left(\dfrac{\$22,000,000}{\$5000}\right)^{(1/50)} - 1 = 18.3\%$, which is much higher than the average annual return for large-company stocks.

69. a. At age 35, the balance in Mitch's account is
$A = \dfrac{\$1000\left[(1+0.05)^{10} - 1\right]}{0.05} = \$12,577.89$. At this point, it is no longer appropriate to use the savings plan formula as Mitch is not making further deposits into the account. Instead, use this balance as the principal in the compound interest formula, and compute the balance 40 years later.
$A = \$12,578(1+0.05)^{40} = \$88,548.20$.

b. At age 75, the balance in Bill's account is
$A = \dfrac{\$1000\left[(1+0.05)^{30} - 1\right]}{0.05} = \$66,438.85$.

c. Mitch deposited $1000 per year for ten years, which is $10,000. Bill deposited $1000 per year for 30 years, which is $30,000, or three times as large as Mitch's total deposits.

d. Answer will vary.

TECHNOLOGY EXERCISES

77. a. FV(0.04/12,12*25,100,0) = $51,412.95
   b. FV(0.08/12,12*25,100,0) = $95,102.64.
      The amount is less than double
   c. FV(0.04/12,12*50,100,0) = $190,935.64.
      The amount is more than double.

79. a. $2.8^{1/4} = 1.293569$
   b. $120^{1/3} = 4.932424$
   c. We need to solve the equation $1850 = 250 \times (1+r)^{15}$. Using algebra, rewrite the equation as $\dfrac{1850}{250} = (1+r)^{15}$.
   Now raise both sides to the reciprocal power of 15, i.e. to the 1/15 power, and then subtract 1 to find $r = \left(\dfrac{1850}{250}\right)^{1/15} - 1 = 0.1427 = 14.27\%$.

UNIT 4D

QUICK QUIZ

1. **a**. Monthly payments go up when the loan principal is larger because you are borrowing more money.

2. **a**. The payment will be higher for a 15-year loan because there is less time to pay off the principal.

3. **b**. A higher APR means more money goes to paying off interest, and this corresponds with a higher payment.

4. **b**. Most of an early payment goes to interest, because the principal is large at the beginning stages of a loan, and thus the interest is also large.

5. **c**. Every year you'll pay $12,000, and thus after ten years, you'll pay $120,000.

6. **c**. Most credit card loans only require that you make a minimum payment each month; there is no specified time in which you must pay off a loan.

7. **c**. A two-point origination fee just means that you must pay 2% of the loan principal in advance, and 2% of $200,000 is $4000.

8. **b**. Add $500 to 1% (one point) of the loan principle of $120,000 to find the advanced payment required.

9. **c**. Refinancing a loan is not a good idea if you've nearly paid the loan off.

10. **a**. A shorter loan always has higher monthly payments (all other things being equal) because there is less time to pay off the principal, and you'll spend less on interest because the principal decreases more quickly.

DOES IT MAKE SENSE?

7. Makes sense. For a typical long-term loan, most of the payments go toward interest for much of the loan term. See Figure 4.8 in the text.

9. Does not make sense. Making only the minimum required payment on a credit card is asking for financial trouble as the interest rates on credit cards are typically exorbitant. You'll end up spending a lot of your money on interest.

11. Makes sense. In the worst-case scenario, the ARM loan will be at 6% in the last year you expect to live in the home, so it's certain you'll save money on the first two years, and you may even save during the last year.

BASIC SKILLS AND CONCEPTS

13. a. The starting principal is $80,000 with an annual interest rate of 7%. You'll make 12 payments each year (one per month) for 20 years, and each payment will be $620.

b. Since you make 12 payments per year for 20 years, you'll make $12 \times 20 = 240$ payments in total, which amounts to $240 \times \$620 = \$148,800$ over the term of the loan.

c. The loan principal is $80,000/$148,800 = 53.8% of the total payment. $148,800 – $80,000 = $68,800, so $68,800/$148,800 = 46.2% of the total payment is interest.

15. a. The monthly payment is

$$PMT = \frac{\$50,000\left(\frac{0.06}{12}\right)}{1-\left(1+\frac{0.06}{12}\right)^{-12\times20}} = \$358.22 \,.$$

b. The total payment is $\$358.22 \times 12 \times 20 = \$85,973$.

c. The percent spent on interest is $\frac{\$85,973-\$50,000}{\$85,973} = 0.418 = 41.8\%$, while

$\frac{\$50,000}{\$85,973} = 0.582 = 58.2\%$ goes toward the principal.

17. a. The monthly payment is

$$PMT = \frac{\$200,000\left(\frac{0.035}{12}\right)}{1-\left(1+\frac{0.035}{12}\right)^{-12\times30}} = \$898.09 \,.$$

b. The total payment is $\$898.09 \times 12 \times 30 = \$323,312$.

c. The percent spent on interest is $\frac{\$323,312-\$200,000}{\$323,312} = 0.381 = 38.1\%$, while

$\frac{\$200,000}{\$323,312} = 0.619 = 61.9\%$ goes toward the principal.

19. a. The monthly payment is

$$PMT = \frac{\$200,000\left(\frac{0.03}{12}\right)}{1-\left(1+\frac{0.03}{12}\right)^{-12\times15}} = \$1381.16 \,.$$

b. The total payment is $\$1381.16 \times 12 \times 15 = \$248,609$.

c. The percent spent on interest is $\frac{\$248,609-\$200,000}{\$248,609} = 0.196 = 19.6\%$, while

$\frac{\$200,000}{\$248,609} = 0.804 = 80.4\%$ goes toward the principal.

21. a. The monthly payment is

$$PMT = \frac{\$10,000\left(\frac{0.08}{12}\right)}{1-\left(1+\frac{0.08}{12}\right)^{-12\times3}} = \$313.36 \,.$$

b. The total payment is $\$313.36 \times 12 \times 3 = \$11,2801$.

c. The percent spent on interest is $\frac{\$11,281-\$10,000}{\$11,281} = 0.114 = 11.4\%$, while

$\frac{\$10,000}{\$11,281} = 0.886 = 88.6\%$ goes toward the principal.

23. a. The monthly payment is

$$PMT = \frac{\$150,000\left(\frac{0.05}{12}\right)}{1-\left(1+\frac{0.05}{12}\right)^{-12\times15}} = \$1186.19 \,.$$

b. The total payment is $\$1186.19 \times 12 \times 15 = \$213,514$.

c. The percent spent on interest is $\frac{\$213,514-\$150,000}{\$213,514} = 0.297 = 29.7\%$, while

$\frac{\$150,000}{\$213,514} = 0.703 = 70.3\%$ goes toward the principal.

25. The monthly payment is

$$PMT = \frac{\$150,000\left(\frac{0.04}{12}\right)}{1-\left(1+\frac{0.04}{12}\right)^{-12\times30}} = \$716.12 \,.$$

25. (continued)

To compute the interest owed at the end of one month, multiply the beginning principal of $150,000 by the monthly interest rate of 4%/12. This gives $150,000(0.04/12) = $500.00. Because the monthly payment is $716.12, the amount that goes toward principal is $716.12 − $500.00 = $216.12. Subtract this from $150,000 to get a new principal of $150,000 − $216.12 = $149,783.88.

For the second month, repeat this process, but begin with a principal of $149,783.88. The results are shown in the following table through the third month − note that rounding errors at each step propagate through the table, which means that our answers may differ slightly from those that a bank would compute.

End of month	Interest	Payment toward principal	New Principal
1	$500.00	$216.12	$149,783.88
2	$499.28	$216.84	$149,567.04
3	$498.56	$217.56	$149,349.48

27. For the 3-year loan at 7% APR, the monthly payment is $PMT = \dfrac{\$12{,}000\left(\dfrac{0.07}{12}\right)}{1-\left(1+\dfrac{0.07}{12}\right)^{-12\times3}} = \$370.53$.

The monthly payment for the 4-year loan is $290.15, and for the 5-year loan, it is $243.32 (the computations being similar to those used for the 3-year loan). Since you can afford only the payments on the 5-year loan, that loan best meets your needs.

29. a. The monthly payment is

$$PMT = \dfrac{\$5000\left(\dfrac{0.18}{12}\right)}{1-\left(1+\dfrac{0.18}{12}\right)^{-12\times1}} = \$458.40.$$

b. Your total payments are $458.40 × 12 = $5500.80.

c. The interest is $500.80/$5500.80 = 9.1% of the total payments.

31. a. The monthly payment is

$$PMT = \dfrac{\$5000\left(\dfrac{0.21}{12}\right)}{1-\left(1+\dfrac{0.21}{12}\right)^{-12\times3}} = \$188.38.$$

b. Your total payments are $188.38 × 36 = $6781.68.

c. The interest is $1781.68/$6781.68 = 26.3% of the total payments.

33. Complete the table as shown in the first month. To find the interest for the second month, compute $1093.00 × 0.015 = $16.40. To find the new balance, compute $1093.00 + $75 + $16.40 − $200 = $984.40. Continue in this fashion to fill out the rest of the table. Note that rounding errors at each step will propagate through the table: a credit card company computing these balances might show slightly different values.

Month	Pmt	Expenses	Interest	New balance
0				$1200.00
1	$200	$75	$18.00	$1093.00
2	$200	$75	$16.40	$984.40
3	$200	$75	$14.77	$874.17
4	$200	$75	$13.11	$762.28
5	$200	$75	$11.43	$648.71
6	$200	$75	$9.73	$533.44
7	$200	$75	$8.00	$416.44
8	$200	$75	$6.25	$297.69
9	$200	$75	$4.47	$177.16
10	$200	$75	$2.66	$54.82

In the 11th month, a partial payment of $54.82 + $54.82 × 0.015 + $75 = $130.64 will pay off the loan.

35. Complete the table as shown in the first month. To find the interest for the second month, compute $179.50 × 0.015 = $2.69. To find next month's balance, compute $179.50 + $150 + $2.69 − $150 = $182.19. Continue in this fashion to fill out the rest of the table. Note that rounding errors at each step will propagate through the table: a credit card company computing these balances might slow slightly different values.

35. (continued)

Month	Pmt	Expenses	Interest	Balance
0				$300.00
1	$300	$175	$4.50	$179.50
2	$150	$150	$2.69	$182.19
3	$400	$350	$2.73	$134.92
4	$500	$450	$2.02	$86.94
5	$0	$100	$1.30	$188.24
6	$100	$100	$2.82	$191.06
7	$200	$150	$2.87	$143.93
8	$100	$80	$2.16	$126.09

37. Under Option 1, the monthly payment is

$$PMT = \frac{\$200,000\left(\frac{0.08}{12}\right)}{1-\left(1+\frac{0.08}{12}\right)^{-12\times30}} = \$1467.53,$$ and the

total cost of the loan is $1467.53 × 12 × 30 = $528,310.80. Under Option 2, the monthly payment is

$$PMT = \frac{\$200,000\left(\frac{0.075}{12}\right)}{1-\left(1+\frac{0.075}{12}\right)^{-12\times15}} = \$1854.02,$$ and the

total cost of the loan is $1854.02 × 12 × 15 = $333,723.60. You'll pay considerably more interest over the term of the loan in Option 1, but you may be able to afford the payments under that option more easily.

39. Under Option 1, the monthly payment is

$$PMT = \frac{\$60,000\left(\frac{0.0715}{12}\right)}{1-\left(1+\frac{0.0715}{12}\right)^{-12\times30}} = \$405.24,$$ and the

total cost of the loan is $405.24 × 12 × 30 = $145,886.40. Under Option 2, the monthly payment is

$$PMT = \frac{\$60,000\left(\frac{0.0675}{12}\right)}{1-\left(1+\frac{0.0675}{12}\right)^{-12\times15}} = \$530.95,$$ and the

total cost of the loan is $530.95 × 12 × 15 = $95,571.00. See Exercise 37 for pros and cons.

41. Under Choice 1, your monthly payment is

$$PMT = \frac{\$120,000\left(\frac{0.04}{12}\right)}{1-\left(1+\frac{0.04}{12}\right)^{-12\times30}} = \$572.90,$$ and the

closing costs are $1200. For Choice 2, the payment is

$$PMT = \frac{\$120,000\left(\frac{0.035}{12}\right)}{1-\left(1+\frac{0.035}{12}\right)^{-12\times30}} = \$538.85,$$ and the

closing costs are $1200 + $120,000(0.02) = $3600. You'll save $572.90 – $538.85 = $34.05 each month with Choice 2, though it will take $3600/($34.05 per month) = 106 months (about 9 years) to recoup the higher closing costs, and thus Choice 2 is the better option only if you intend to keep the house for at least 9 years.

43. Under Choice 1, your monthly payment is

$$PMT = \frac{\$120,000\left(\frac{0.045}{12}\right)}{1-\left(1+\frac{0.045}{12}\right)^{-12\times30}} = \$608.02,$$ and the

closing costs are $1200 + $120,000(0.01) = $2400. For Choice 2, the payment is

$$PMT = \frac{\$120,000\left(\frac{0.0425}{12}\right)}{1-\left(1+\frac{0.0425}{12}\right)^{-12\times30}} = \$590.33,$$ and the

closing costs are $1200 + $120,000(0.03) = $4800. You'll save $608.02 – $590.33 = $17.69 each month with Choice 2, though it will take $2400/($17.29 per month) = 139 months (12 years) to recoup the higher closing costs, and thus Choice 2 is the better option only if you intend to keep the house for at more than 11 years.

45. a. The monthly payment is

$$PMT = \frac{\$30,000\left(\frac{0.09}{12}\right)}{1-\left(1+\frac{0.09}{12}\right)^{-12\times20}} = \$269.92.$$

b. The monthly payment is

$$PMT = \frac{\$30,000\left(\frac{0.09}{12}\right)}{1-\left(1+\frac{0.09}{12}\right)^{-12\times10}} = \$380.03.$$

45. (continued)

 c. If you pay the loan off in 20 years, the total payments will be $269.92 × 12 × 20 = $64,780.80. If you pay it off in 10 years, the total payments will be $380.03 × 12 × 10 = $45,603.60.

47. Following example 9 in the text, the interest on the $150,000 loan in the first year will be approximately 4.5% × $150,000 = $6750, which means the monthly payment will be about $6750/12 = $562.50. With the 3% ARM, the interest will be approximately 3% × $150,000 = $4500, and thus the monthly payment will be about $4500/12 = $375. You'll save $562.50 − $375 = $187.50 each month with the ARM. In the third year, the rate on the ARM will be 2.0 percentage points higher than the fixed rate loan, and thus the yearly interest payments will differ by $150,000 × 0.02 = $3000, which amounts to $250 per month.

FURTHER APPLICATIONS

49. Solve $\$500 = \dfrac{P\left(\dfrac{0.0375}{12}\right)}{1-\left(1+\dfrac{0.0375}{12}\right)^{-12\times30}}$ for P by computing the values in the numerator and denominator on the right, and isolating P. This results in $P = \$107,964$ (rounded to the nearest dollar), and it represents the largest loan you can afford with monthly payments of $500. The loan will pay for 80% of the house you wish to buy (you must come up with a 20% down payment). Thus 80% × (*house price*) = $107,964, which means *house price* = $107,964 /0.80 = $134,956. You can afford a house that will cost about $134,956.

51. a. The monthly payment for the first loan is

$$PMT = \frac{\$10,000\left(\dfrac{0.065}{12}\right)}{1-\left(1+\dfrac{0.065}{12}\right)^{-12\times15}} = \$87.11$$

The payments for the other two loans are calculated similarly; you should get $116.30, $148.38, respectively.

b. For the first loan, you'll pay $87.11 × 12 × 15 = $15,679.80 over its term. The total payments for the other loans are computed in a similar fashion – you should get $27,912 and $17,805, respectively. Add them together to get $61,397.

c. The monthly payment for the consolidated loan will be $PMT = \dfrac{\$37,500\left(\dfrac{0.065}{12}\right)}{1-\left(1+\dfrac{0.065}{12}\right)^{-12\times15}} = \$326.67$, and the total payment over 15 years will be $58,801. You'll pay about $2596 less for the consolidated loan ($61,397 − $58,801= $2596), and your monthly payments will be lower for the first ten years ($326.67 versus $87.11 + $116.30 + $148.38 = $351.79), but higher once the ten year loan is paid off. It would probably be worth it to consolidate the loans. It also means you'll have to keep track of only one loan instead of three, though with automatic withdrawals offered by most banks, this isn't as much of an issue today as it might have been 15 years ago.

53. a. The annual ($n = 1$) payment is $8718.46; the monthly ($n = 12$) payment is $716.43, the bi-weekly ($n = 26$) payment is $330.43, and the weekly ($n = 52$) payment is $165.17. The weekly payment is computed below; the others are done in a similar fashion.

$$PMT = \frac{\$100,000\left(\dfrac{0.06}{52}\right)}{1-\left(1+\dfrac{0.06}{52}\right)^{-52\times20}} = \$165.17 .$$

b. The total payout for each scenario is as follows: $174,369.20 ($n = 1$); $171,943.20 ($n = 12$); $171,823.60 ($n = 26$); and $171,776.80. The total payout when $n = 52$ is shown below; the others are done in a similar fashion.

$$\frac{\$165.17}{\text{week}} \times \frac{52\text{ weeks}}{\text{year}} \times 20\text{ years} = \$171,776.80 .$$

c. The total payout decreases as n increases.

55. Answers will vary.

TECHNOLOGY EXERCISES

63. a. PMT(0.09/12,12*10,7500) = $95.01

 b. PMT(0.09/12,12*20,7500) = $67.48.

The payment decreases

 c. PMT(0.045/12,12*10,7500) = $77.73.

The payment decreases.

UNIT 4E

QUICK QUIZ

1. **a**. Gross income is defined as the total of all income you receive.

2. **c**. The first portion of your income is taxed at 10% (the portion being determined by your filing status – see Table 4.9), the next at 15%, and only the last portion is taxed at 25%.

3. **a**. A tax credit reduces your tax bill by the dollar amount of the credit, so your tax bill will be reduced by $1000.

4. **b**. A tax deduction reduces your taxable income, so if you have a deduction of $1000, and you are in the 15% tax bracket, your bill will be reduced by 15% of $1000, or $150. This answer assumes you are far enough into the 15% tax bracket so that reducing your taxable income by $1000 means you remain in the 15% bracket.

5. **b**. You can claim deductions of $5000 + $2000 = $7000 as long as you itemize deductions.

6. **a**. If you chose to itemize your deductions, the only thing you can claim is the $1000 contribution, and as this is less than your standard deduction of $6100, it won't give relief to your tax bill (in fact, you'd pay more tax it you were to itemize deductions in this situation – take the standard deduction).

7. **c**. FICA taxes are taxes levied on income from wages (and tips) to fund the Social Security and Medicare programs.

8. **a**. Joe will pay 7.65% of his income for FICA taxes. Kim pays 7.65% of only her first $113,700; she pays 1.45% of her remaining salary, for a total FICA tax of 0.0765($113,700) + 0.0145($150,000 – $113,700) = $9224.40, which is 6.1% of her income. David pays nothing at all in FICA taxes because income from capital gains is not subject to FICA taxes.

9. **b**. Assuming all are of the same filing status, Jerome pays the most because his FICA taxes will be highest, followed by Jenny, and then Jacqueline.

10. **c**. Taxes on money deposited into tax-deferred accounts are *deferred* until a later date: you don't have to pay taxes on that money now, but you will when you withdraw the money in later years.

DOES IT MAKE SENSE?

11. Does not make sense. We know nothing of the rest of the picture for these two individuals – they may have very different deductions, and thus different tax bills.

13. Makes sense. You can deduct interest paid on the mortgage of a home if you itemize deductions, and for many people, the mortgage interest deduction is the largest of all deductions.

15. Makes sense (in some scenarios). Bob and Sue might be hit with the "marriage penalty" when they file taxes, and if they postpone their wedding until the new year, they could avoid the penalty for at least the previous tax year.

17. Makes sense, because at $10,000 your personal exemption and standard deduction would give you $0 taxable income, but being self-employed means you are still subject to the 15.3% self-employment FICA tax.

BASIC SKILLS AND CONCEPTS

19. Antonio's gross income was $47,200 + $2400 = $49,600. His AGI was $49,600 – $3500 = $46,100. His taxable income was $46,100 – $3900 – $6100 = $36,100.

21. Isabella's gross income was $88,750 + $4900 = $93,650. Her AGI was $93,650 – $6200 = $87,450. Her taxable income was $87,450 – $3900 – $9050 = $74,500.

23. Your itemized deductions total to $8600 + $2700 + $645 = $11,945, and since this is smaller than the standard deduction of $12,200, you should claim the standard deduction and not itemize your deductions.

25. Suzanne's gross income is $33,200 + $350 = $33,550, her AGI is $33,550 – $500 = $33,050, and her taxable income is $33,050 – $3900 – $6100 = $23,050. She should claim the standard deduction because it is much higher than her itemized deduction of $450.

27. Wanda's gross income was $33,400 + $500 = $33,900, her AGI was the same, and her taxable income was $33,900 – $3900 × 3 – $6100 = $16,100. She should claim the standard deduction because it is higher than her itemized deduction of $1500.

29. Gene owes 10% × ($8925) + 15% × ($35,400 – $8925) = $ 4863.75 (Taxes are rounded to the nearest dollar, so the tax owed is $4864.)

31. Bobbi owes 10% × ($8925) + 15% × ($36,250 – $8925) + 25% × ($73,200 – $36,250) + 28% × ($77,300 – $73,200) = $15,256.75. (Taxes are rounded to the nearest dollar, so the tax owed is $15,257.)

33. Paul owes 10% × ($12,750) + 15% × ($48,600 – $12,750) + 25% × ($89,300 – $48,600) – $1000 = $15,827.50. (Taxes are rounded to the nearest dollar, so the tax owed is $15,828.)

35. Winona and Jim owe 10% × ($17,850) + 15% × ($72,500 – $17,850) + 25% × ($105,500 – $72,500) – $2000 = $16,232.50. (Taxes are rounded to the nearest dollar, so the tax owed is $16,233.)

37. Their bill will be reduced by $500.

39. Her taxes will not be affected because she is claiming the standard deduction, and a $1000 charitable contribution must be itemized.

41. His tax bill will be reduced by 28% of $1000, or $280, provided he is far enough into the 28% bracket that his $1000 contribution does not drop him into the lower bracket.

43. The $1800 per month of your payment that goes toward interest would be deductible, which means you'd save 33% × $1800 = $594 each month in taxes. Thus your $2000 mortgage payment would effectively be only $2000 – $594 = $1406, which is less than your rent payment of $1600. Buying the house is cheaper. This solution assumes that the $1800 × 12 = $21,600 interest deduction would not drop you into the 28% bracket (if it did, the solution would be more complicated).

45. Maria saves 33% of $10,000, or $3300 in taxes, which means the true cost of her mortgage interest is $6700. Steve saves 15% of $10,000, or $1500 in taxes, which means the true cost of his mortgage interest is $8500. This solution assumes their deductions do not drop them into the lower tax bracket.

47. Luis will pay 7.65% × $28,000 = $2142 in FICA taxes. His taxable income is $28,000 – $3900 – $6100 = $18,000. He will pay 10% × ($8925) + 15% × ($18,000– $8925) = $2254 in income taxes, and thus his total tax bill will be $2142 + $2254 = $4396. This is $4396/$28,000 = 15.7% of his gross income, so his effective tax rate is 15.7%.

49. Jack will pay 7.65% × $44,800 = $3427 in FICA taxes. His gross income is $44,800 + $1250 = $46,050, and his taxable income is $46,050 – $3900 – $6100 = $36,050. He will pay 10% × ($8925) + 15% × ($36,050 – $8925) = $4961 in income taxes, and thus his total tax bill will be $34270 + $4961 = $8388. This is $8388/$46,050 =

18.2% of his gross income, so his effective tax rate is 18.2%.

51. Brittany will pay 7.65% × $48,200 = $3687 in FICA taxes. Her taxable income is $48,200 – $3900 – $6100 = $38,200. She will pay 10% × ($8925) + 15% × ($36,250 – $8925) + 25% × ($38,200 – $36,250) = $5479 in income taxes, and thus her total tax bill will be $3687 + $5479 = $9166. This is $9166/$48,200 = 19.0% of her gross income, so her effective tax rate is 19.0%.

53. Pierre will pay 7.65% × ($113,700) + 1.45% × ($120,000 – $113,700) = $8789 in FICA taxes, and because his taxable income is $120,000 – $3900 – $6100 = $110,000, he will pay 10% × ($8925) + 15% × ($36,250 – $8925) + 25% × ($87,850 – $36,250) + 28% × ($110,000 – $87,850) = $24,093 in income taxes. His total tax bill is $8789 + $24,093 = $32,882, so his overall tax rate is $32882/$120,000 = 27.4%. Like Pierre, Katarina's taxable income is $110,000, but she is taxed at the special rates for dividends and long-term capital gains. She will pay nothing in FICA taxes, and her income taxes will be 0% × ($36,250) + 15% × ($110,000 – $36,250) = $11,063. Her overall tax rate is $11,063/$120,000 = 9.2%.

55. Because you are in the 15% tax bracket, every time you make a $400 contribution to your tax-deferred savings plan, you save 15% × $400 = $60 in taxes. Thus your take-home pay is reduced not by $400, but by $340.

57. Because you are in the 25% tax bracket, every time you make an $800 contribution to your tax-deferred savings plan, you save 25% × $800 = $200 in taxes. Thus your take-home pay is reduced not by $800, but by $600.

FURTHER APPLICATIONS

59. Gabriella's taxable income is $96,400 – $3900 – $6100 = $86,400, so her income tax is 10% × ($8925) + 15% × ($36,250 – $8925) + 25% × ($86,400 – $36,250) = $17,528.75. Roberto's taxable income is $82,600 – $3900 – $6100 = $72,600, so his income tax is 10% × ($8925) + 15% × ($36,250 – $8925) + 25% × ($72,600 – $36,250) = $14,078.75. If they delay their marriage, they will pay $17,528.75 + $14,078.75 = $31,607.50 in taxes. If they marry, their combined AGI will be $96,400 + $82,600 = $179,000, and their taxable income will be $179,000 – $3900 × 2 – $12,200 = $159,000. Their income taxes will be 10% × ($17,850) + 15% × ($72,500 – $17,850) + 25% × ($146,400 – $72,500) + 28%

59. (continued)

    $\times$ ($159,000 − $146,400) = $31,985.50, which is slightly higher than the taxes they would pay as individuals, and thus they will face a marriage penalty.

61. Steve's taxable income is $185,000 − $3900 − $6100 = $175,000, so his income tax is 10% $\times$ ($8925) + 15% $\times$ ($36,250 − $8925) + 25% $\times$ ($87,850 − $36,250) + 28% $\times$ ($175,000 − $87,850) = $42,293.25. Mia's tax is, of course, the same, so if they delay their marriage, they will pay $84,587. If they marry, their combined AGI will be $185,000 $\times$ 2 = $370,000, and their taxable income will be $370,000 − $3900 $\times$ 2 − $12,200 = $350,000. Their income taxes will be 10% $\times$ ($17,850) + 15% $\times$ ($72,500 − $17,850) + 25% $\times$ ($146,400 − $72,500) + 28% $\times$ ($223,050 − $146,400) + 33% $\times$ ($350,000 − $223,050) = $91,813, which is more than the taxes they would pay as individuals, and thus they will face a marriage penalty.

63. a. Deirdre will pay 7.65% $\times$ ($90,000) = $6885 in FICA taxes. Robert will also pay $6885 in FICA taxes. Jessica and Frank will pay 2 $\times$ ($6885) = $13,770 in FICA taxes.

    b. From Example 3, Deirdre will pay $15,929 + $6885 = $22,814, Robert will pay $11,840 + $6885 = $18,725, and Jessica and Frank will pay $32,266 + $13,770 = $46,036.

    c. Deirdre's overall tax rate:

    $$\frac{\$22,814}{\$90,000} = 25.3\%$$

    Robert's overall tax rate:

    $$\frac{\$18,725}{\$90,000} = 20.8\%$$

    Jessica and Frank's overall tax rate:

    $$\frac{\$46,036}{\$180,000} = 25.6\%$$

    d. Serena had the lowest overall tax rate, following by Robert, Deirdre, and Jessica and Frank, in ascending order.

    e. Serena receives by far the greatest tax break because all of her income is from investments.

65. – 67. Answers will vary.

UNIT 4F

QUICK QUIZ

1. **b**. Bigprofit.com had more outlays (expenses) than receipts (income) by $1 million, so it ran a deficit for 2009 of $1 million. Its debt is the total amount owed to lenders over the years; this is $7 million.

2. **c**. The federal debt is about $17 trillion dollars, and when divided evenly among all 315 million U.S. citizens, it comes to about $54,000 per person.

3. **c**. Tax revenues are below average.

4. **b**. Discretionary outlays differ from mandatory outlays in that Congress passes a budget every year that spells out where the government will spend its discretionary funds.

5. **a**. Interest on the debt must be paid so that the government does not default on its loans, and Medicare is an entitlement program that is part of mandatory spending. On the other hand, all money spent on defense is considered discretionary spending.

6. **a**. Mandatory expenses include money spent on Social Security, Medicare, interest on the debt, government pensions, and Medicaid, and together these programs constitute almost 60% of the budget (see Figure 4.14).

7. **b**. The $100 billion is supposed to be deposited into an account used to pay off future Social Security recipients, but instead, the government spends the money at hand, and essentially writes IOUs to itself, and places these in the account.

8. **b**. Publicly held debt is the money the government owes to those who have purchased Treasury bills, notes, and bonds.

9. **b**. The gross debt is the sum of all of the money that the government owes to individuals who have purchased bonds, and the money it owes to itself.

10. **c**. The amount of money set aside for education grants is miniscule in comparison to the billions of dollars the government will be required to spend on Social Security in 2030.

DOES IT MAKE SENSE?

9. Makes sense. Each U.S. citizen would have to spend about $43,000 to retire the federal debt, and it's certainly true that this exceeds the value of most new cars.

11. Does not make sense. The financial health of the government is very much dependent upon what happens with Social Security and the FICA taxes collected to fund it. The term "off-budget" is nothing more than a label that separates the Social Security program from the rest of the budget.

13. Does not make sense. There is no guarantee the government will have the funds to pay for social security 40 years from now.

BASIC SKILLS AND CONCEPTS

15. a. Your receipts are $38,000, and your outlays are $12,000 + $6000 + $1200 + $8500 = $27,700, so you have a surplus.

    b. A 3% raise means your receipts will be $38,000 + 3% × $38,000 = $39,140, and your outlays will be $27,700 + $8500 = $36,200, so you'll still have a surplus, though it will be smaller.

    c. Your receipts will be $39,140 (see part **b**), and your outlays will be $27,700 × 1.01 + $7500 = $35,477, so you will be able to afford it.

17. a. The interest on the debt will be 8.2% × $773,000 = $63,000 (rounded to the nearest thousand dollars).

    b. The total outlays are $600,000 + $200,000 + $250,000 + $63,000 = $1,113,000, so the year-end deficit is $1,050,000 − $1,113,000 = −$63,000, and the year-end accumulated debt is −$773,000 − $63,000 = −$836,000.

    c. The interest on the debt in 2015 will be 8.2% × $836,000 = $69,000 (rounded to the nearest thousand dollars).

    d. The total outlays are $600,000 + $200,000 + $69,000 = $869,000, so the year-end surplus is $1,100,000 − $869,000 = $231,000, and the year-end accumulated debt is −$836,000 + $231,000 = −$605,000, assuming all surplus is devoted to paying down the debt.

19. $$\frac{\$1.61\times10^{13}}{170\times10^{6}\text{ workers}}=\frac{\$94,706}{\text{worker}}$$

21. 2000 Surplus: $\dfrac{\$236\times10^{9}}{\$9821\times10^{9}}=2.4\%$

    2000 Debt: $\dfrac{\$5500\times10^{9}}{\$9821\times10^{9}}=56.0\%$

    2009 Deficit: $\dfrac{-\$1412\times10^{9}}{\$13,937\times10^{9}}=10.1\%$

    2009 Debt: $\dfrac{\$12,000\times10^{9}}{\$13,937\times10^{9}}=86.1\%$

23. 2009 Deficit: $\dfrac{-\$1412\times10^{9}}{\$13,937\times10^{9}}=10.1\%$

    2017 Deficit: $\dfrac{-\$480\times10^{9}}{\$20,000\times10^{9}}=2.4\%$

    This is a change of (2.4% − 10.1%)/10.1% = −76%.

25. At 2.1%, the interest payment would be 2.1% × $20 trillion = $420 billion. With the 0.5% increase, the interest payment would be 2.6% × $20 trillion = $520 billion. The increase in the interest rate is (2.6% − 2.1%) / 2.1% = 24%, which would be the same increase as the interest payment.

27. Revenue from individual income taxes was 47% × $2.9 trillion = $1.36 trillion. Revenue from social insurance was 34% × $2.9 trillion = $0.986 trillion or $986 billion.

29. Spending would increase by 1.6% × $3.5 trillion = $0.056 trillion or $56 billion, while revenue would increase by 2% × $2.9 trillion = $0.058 trillion or $58 billion, so it would be possible.

31. Spending on Social Security and Medicare increases by 39% − 33% = 6%, which is 6% × $3.5 trillion = $0.21 trillion or $210 billion. Non-discretionary spending would decrease by the same values.

33. Because *unified net income − off-budget net income = on-budget net income*, we have $40 billion − $180 billion = −$140 billion. In other words, despite the fact that the government proclaimed a surplus of $40 billion for this year, it actually ran a deficit of $140 billion due to the fact that the money earmarked for the Social Security trust fund was never deposited there (it was spent on other programs).

35. The government could cut discretionary spending, it could borrow money by issuing Treasury notes, or it could raise taxes (of course, a combination of these three would also be an option).

FURTHER APPLICATIONS

37. It would take about $1.7×10^{13}

$$\times\frac{1\text{ s}}{\$1}\times\frac{1\text{ hr}}{3600\text{ s}}\times\frac{1\text{ d}}{24\text{ hr}}\times\frac{1\text{ yr}}{365\text{ d}}\approx539{,}00\text{ years.}$$

39. In ten years, the debt will be

$$\$1.7 \times 10^{13}(1+0.01)^{10} = \$1.88 \times 10^{13},$$

or about $18.8 trillion. In 50 years, it will be

$$\$1.7 \times 10^{13}(1+0.01)^{50} = \$2.8 \times 10^{13},$$

or about $28 trillion.

41. In 2017, federal revenue would be 154% × $2.45 trillion = $3.773 trillion, and federal spending would be 120% × $3.54 trillion = $4.248 trillion. This would result in a deficit in 2017 of $3.773 trillion − $3.248 trillion = −$0.475 trillion. In 2012, the federal deficit was $2.45 trillion − $3.54 trillion = −$1.09 trillion, so the deficit will have decreased.

43. $PMT = \dfrac{\$12.0 \times 10^{12}\left(\dfrac{0.04}{1}\right)}{1-\left(1+\dfrac{0.04}{1}\right)^{-1 \times 10}} = \$1.48 \times 10^{12}$, which

is $1.48 trillion per year.

45. $\$17 \times 10^{12} \times \dfrac{1 \text{ wk}}{\$160 \times 10^{6}} \times \dfrac{1 \text{ yr}}{52 \text{ wk}} = 2043 \text{ years}$.

UNIT 5A

QUICK QUIZ

1. **a.** The population is the complete set of people or things that are being studied.

2. **c.** Those who donated money to the governor's campaign are much more likely to approve of his job.

3. **a.** A representative sample is one where the characteristics of the sample match those of the population.

4. **b.** Those who do not receive the treatment (in this case, a cash incentive) are in the control group. Note that the teacher could also randomly select a third group of students who don't even know they are part of the experiment – she could study their attendance rates, and this would be another control group.

5. **c.** The experiment is not blind, because the participants know whether they are receiving money or not, and the teacher knows to which group each student belongs.

6. **a.** The purpose of a placebo is to control for psychological effects that go along with being a participant of the study.

7. **c.** A placebo effect results when participants of the study improve because they think they are receiving the treatment.

8. **c.** A single-blind experiment is one where the participants do not know whether they are part of the treatment group or the control group, but the person conducting the experiment does know.

9. **b.** With a margin of error of 3%, both Poll X and Poll Y have predictions that overlap (in that Poll X predicts Powell will receive 46% to 52% of the vote, and Poll Y predicts he will receive 50% to 56% of the vote), and so they are consistent in their predictions.

10. **b.** The confidence interval is simply the results of the poll plus-or-minus the margin of error.

DOES IT MAKE SENSE?

9. Does not make sense. A sample is a subset of the population, and thus it cannot exceed the size of the population.

11. Does not make sense. The control group should no treatment, not a different treatment than that given to the treatment group.

13. Makes sense. The poll makes a prediction about the outcome of the election, and though we can place a reasonable level of confidence in the results of the poll, it cannot tell the future for us.

BASIC SKILLS AND CONCEPTS

15. The population is all Americans, and the sample was the set of 1001 Americans surveyed. The population parameters are the opinions of Americans on Iran, and the sample statistics consist of the opinions of those who were surveyed.

17. The population is the set of all Americans, and the sample is the set of 998 people who were surveyed. The population parameter is the percent Americans who believe there has been progress in finding a cure for cancer in the last 30 years, and the sample statistic is the 60% of those surveyed that believe there has been progress in finding a cure for cancer in the last 30 years.

19. The population is the set of Americans, and the sample is the set of 1027 Americans surveyed. The population parameter is the percent of those who believe there should be an investigation of anti-terror tactics used during the Bush administration, while the sample statistic is the 62% of those surveyed who believe there should be an investigation of anti-terror tactics used during the Bush administration.

21. *Step 1*: The goal is to determine the average number of hours spent per week talking on cell phones among a population of ninth graders. *Step 2*: Randomly select (from a list of all students at the school) a sample from the population. *Step 3*: Determine cell phone use, in hours per week, from those in the sample, either with an interview, a survey, or a device that measures more precisely when the cell phone is in use. *Step 4*: Conclude that cell phone use among all ninth-graders is similar to cell phone use among the sample. *Step 5*: Assess the results and formulate a conclusion.

23. *Step 1*: The population of this study is American male college students, and the goal is to determine the percentage of the population that plays chess at least once per week. *Step 2*: Choose a representative sample of male college students. *Step 3*: Survey students and determine the percentage of those in the sample who play chess at least once a week. *Step 4*: Infer the percentage

23. (continued)

    of the population who play chess from the results of the study. *Step 5*: Assess the results and draw conclusions.

25. *Step 1*: The population in the study is all batteries in a particular model of laptop, and the goal is to find the average time to failure of those batteries. *Step 2*: Select a representative sample of laptops. *Step 3*: Formulate a method to determine lifetimes of batteries, and compute the average lifetime. *Step 4*: Infer the average lifetime found in *Step 3* is the same for all batteries. *Step 5*: Assess the results and formulate a conclusion.

27. The first 100 first-year students met in the student union is the most representative sample for first-year students at a particular school because they are a more random sample of students at that school. The other samples are not likely to be representative since they would be biased towards subgroups within the population.

29. This is an example of stratified sampling as the sample of taxpayers is broken in several categories (or strata).

31. This is an example of stratified sampling as the participants are divided into groups based on age.

33. This is a simple random sampling because the people are selected at random.

35. This is an observational study, with a case-control element, where the cases are those who have a tendency to lie, and the controls are those who do not lie.

37. This is an experiment where the treated group consists of participants who received treatment with magnets, and the control group consists of participants who received treatment with non-magnets. Double-blinding would be necessary.

39. This is an experiment where the treated group consists of patients who received ginkgo biloba extract, and the control group consists of those who received a placebo. Double-blinding would be necessary.

41. An experiment would answer this question best. Grow two groups of tomatoes where one group of tomatoes is treated with *Fortify!* and the second is not, keeping all other factors the same.

43. An observational study would be sufficient to determine which National Football League has the lineman with the greatest average weight.

45. An experiment would answer this question best. Recruit people to participate in the study, and ask one group to take a multi-vitamin every day, while the other group receives a placebo. Observe the rate of strokes among each group to determine the effect of multi-vitamins.

47. The confidence interval is 50.5% to 55.5%, and because there are only two candidates, it is likely (though not guaranteed) the Republican will win, so a victory party is called for.

49. The confidence interval for the most recent data shows that 43% to 49% of Americans support legalized abortion, so 51% to 57% oppose legalized abortion, so you can claim that a majority of Americans oppose legalized abortion.

FURTHER APPLICATIONS

51. Because a significantly larger percentage of those who received the drug showed improvement compared to those who did not receive the drug, there is good evidence that the treatment is effective (or at least more effective than no treatment at all).

53. Because a significantly larger percentage of those who received the drug showed improvement compared to those who did not receive the drug, there is good evidence that the treatment is effective.

55. a. The population is all North American asymptomatic HIV patients who have undergone drug treatment. The population parameters are the survival rates and times at which treatment began.

    b. The sample is the 17,517 asymptomatic North American patients with HIV. The sample statistics are the survival rates and times at which treatment began for those in sample.

    c. This is on observational study.

    d. The actual points in the progression of the disease

57. a. The population is all people over 60. The population parameters are the test scores for all elderly people, categorized by whether or not each individual was given a suggestion.

    b. The sample is 100 people in the 60–70 and 71–82 age categories. The sample statistics are the test scores categorized by whether or not each individual was given a suggestion.

    c. This is an experiment.

    d. How the sample was selected and how cognitive performance was measured

59. a. The population is all adolescents in 36 states. The population parameters are the percentages of students in each state who became regular smokers.

    b. The sample is 16,000 adolescents in 36 states. The sample statistics are the percentages of students in the samples in each state who became regular smokers.

    c. This is on observational study.

    d. How the sample was selected and how many 10th-graders were in the sample

UNIT 5B

QUICK QUIZ

1. **c**. When the wrong technique is used in a statistical study, one should put little faith in the results.

2. **b**. While it is possible that an oil company can carry out legitimate and worthy research on an environmental problem it caused, the opportunity (and temptation) for bias to be introduced should make you wary of the results.

3. **a**. The fact that the researchers interviewed only those living in dormitories means the sample chosen may not be representative of the population of all freshmen, some of whom don't live on campus.

4. **b**. The poll that determines the winner suffers from participation bias, as those who vote choose to do so. Also true is that those who vote almost certainly watch the show, while the vast majority of Americans do not watch it, nor do they care who wins.

5. **b**. The quantities that a statistical study attempts to measure are called variables of interest, and this study is measuring the weights of 6-year-olds.

6. **a**. The quantities that a statistical study attempts to measure are called variables of interest, and this survey is measuring the number of visits to the dentist.

7. **c**. It's reasonable to assume that people who use sunscreen do so because they spend time in the sun, and this can lead to sunburns.

8. **c**. The availability error is best avoided by carefully choosing the order in which answers to a survey are presented, and in this case, it's best to switch the order for half of the people being polled.

TECHNOLOGY EXERCISES

69. a. Answers will vary

    b. Answers will vary.

    c. The average of a large number of random numbers should be near one-half.

9. **b**. A self-selected survey suffers from participation bias, where people make the choice to participate (that is, they select themselves, rather than being randomly selected by a polling company).

10. **b**. Carefully conducted statistical studies give us good information about that which is being studied, but it's always possible for unforeseen problems (such as confounding variables) to arise, and the "95% confidence" that was spoken of in Unit 5A (and expanded upon in Unit 6D) can always come into play.

DOES IT MAKE SENSE?

5. Does not make sense. A TV survey with phone-in responses suffers from participation bias, whereas a survey carried out by professional pollsters more likely than not will give better results.

7. Does not make sense. Statistical studies can rarely (if ever) make the claim that the results are proven beyond all doubt.

BASIC SKILLS AND CONCEPTS

9. You should not put much faith in the study because the sample is not likely to be representative of the population (SATs are taken before the teachers go to college, which will not be a good indication of the teachers' academic preparation.

11. A study done by a conservative group to assess a new Democratic spending plan could easily include biases, so you should doubt the results.

13. You should doubt a poll that suffers from participation bias, as this call-in poll does.

15. It is very difficult to quantify or even define happiness, and thus this study is essentially meaningless (and you should not trust it, unless you believe happiness was correctly defined and measured).

17. You should doubt the results of this study because there is no control group (riders who do not use helmets) to which results might be compared.

19. You should doubt the results of this study because it uses self reporting, which is not always accurate (especially in cases of sensitive issues, like alcohol abuse).

21. This is a reasonable claim, especially if the effect of inflation is taken into account.

23. Depending on the reliability of the president's evidence, this could be a reasonable claim. You should look for more research of the subject.

25. The Chamber of Commerce would have no reason to distort its data, so the claim is believable.

FURTHER APPLICATIONS

27. The survey will suffer from a selection bias in that this group of shoppers is not representative of the entire population of people with colds.

29. A potential selection bias exists in this survey because the people who vote from 7:00 to 7:30 a.m. may not be a representative sample of the voting population (those in the sample are more likely to be working people who vote on the way to their jobs).

31. This survey will suffer from a selection bias due to the fact that the National Guard is charged with protecting the borders, and thus their views on immigration won't be representative of the entire country.

33. Planned Parenthood members have different views on the issue of contraceptives for high school students than the population of American adults (that is, they are not a representative sample), and thus this survey is influenced by the selection bias.

35. Answers will vary.

37. There is no mention of how the respondents were selected or what questions was asked that had 60% of adults avoiding visits to the dentist because of fear.

39. There is no mention of how the respondents were selected, how many responded, or how the quality of the restaurants was measured.

41. We don't know what question was asked to obtain the responses (did the question just ask for a favorite vegetable, or were potatoes one of a list of options?) We also don't know how the sample was selected.

43. A headline like this implies "illegal drugs" to most readers, and the government study includes smoking and drinking (both of which are legal and largely acceptable practices in many segments of society). The study is based upon top movie rentals, and yet the headline implies all movies. It's a misleading headline.

45. The questions have a different population. The population for the first question is all people who have dated on the Internet. The population for the second question is all married people.

UNIT 5C

QUICK QUIZ

1. **b**. The relative frequency of a category expresses its frequency as a fraction or percentage of the total, and thus is $25/100 = 0.25$.

2. **c**. We would need to know the number of A's assigned before being able to compute the cumulative frequency.

3. **b**. Qualitative data describe non-numerical categories.

4. **c**. The sizes of the wedges are determined by the percentage of the data that corresponds to a particular category, and this is the same as the relative frequency for that category.

5. **a**. Depending on how you want to display the data, you could actually use any of the three options listed. However, a line chart is used for numerical data, which means you would have to suppress the names of the tourist attractions (your categories would become bins, such as "0-2 million"). For a pie chart, you would need to compute the relative frequencies of the data, and while that's not terribly difficult, it requires an extra step. It would be most appropriate to use a bar graph, with the various tourist attractions as categories (e.g. Disney World, Yosemite National Park, etc.), and the annual number of visitors plotted on the y-axis.

6. **c**. The ten tourist attractions are the categories for a bar graph, and these belong on the horizontal axis.

7. **a**. With 100 data points, each precise to the nearest 0.001, it would be best to bin the data into several categories.

8. **c**. A line chart is often used to represent time-series data.

9. **c**. A histogram is a bar graph for quantitative data categories.

10. **b**. For each dot in a line chart, the horizontal position is the center of the bin it represents.

DOES IT MAKE SENSE?

7. Does not make sense. A frequency table has two columns, one of which lists the frequency of the various categories.

9. Does not make sense. The width of the bars in a bar chart is not important (though they should all be of the *same* width in a histogram, and even for those bar charts that are not histograms, even-width bars look better).

11. Does not make sense. The position of the wedges in a pie chart is immaterial.

13. Does not make sense. Bar charts can be used for qualitative data: the categories are the qualities of concern, and the vertical axis shows the frequency of each category (or the relative frequencies).

BASIC SKILLS AND CONCEPTS

15.

Grade	Freq.	Rel. freq.	Cum. freq.
A	2	0.10	2
B	5	0.25	7
C	8	0.40	15
D	3	0.15	18
F	2	0.10	20
Total	20	1.00	20

17. Birth month is qualitative data, even if listed numerically.

19. The yes/no responses on the ballot are qualitative data because it describes a non-numerical quality of opinion.

21. Flavors of bagels are qualitative data because they are non-numerical choices in the poll.

23. Annual salaries are quantitative data because it is a numerical measurement of money.

25.

Bin	Freq.	Rel. freq.	Cum. freq.
95–99	3	0.15	3
90–94	2	0.10	5
85–89	3	0.15	8
80–84	2	0.10	10
75–79	4	0.20	14
70–74	1	0.05	15
65–69	3	0.15	18
60–64	2	0.10	20
Total	20	1.00	20

27.

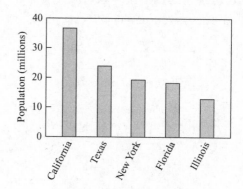

29.

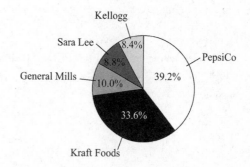

31.

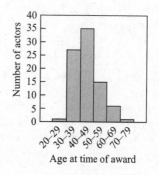

33.

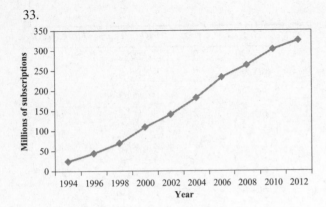

FURTHER APPLICATIONS

35. a. The ages are quantitative categories.

 b.

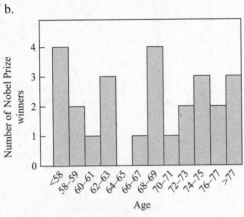

 c. People who win the Nobel Prize (in literature) do so rather late in their career.

37. a. The data are quantitative.

 b.

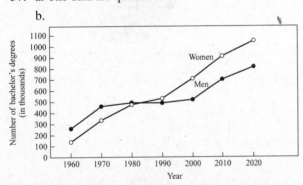

 c. The number of bachelor's degrees conferred on both men and women have been increasing and more women than men now earn bachelor's degrees.

39. a. The data are quantitative.

 b.

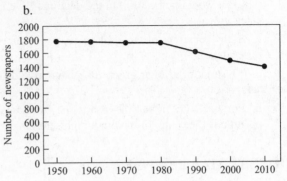

 c. The number of newspapers is declining.

41. a. The religions are qualitative categories.

 b.

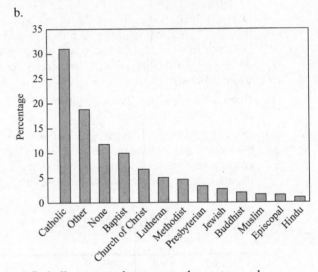

 c. Catholics outnumber every other category by a significant percentage.

43. a. The years are quantitative categories.

 b.

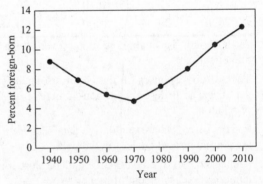

 c. The percentage decreased until 1970 and began increasing after that.

45. a.

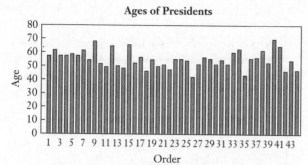

Ages of Presidents

b. The presidents that could have said that they were the *youngest* president (or the same age in years as the youngest) at the time they took office are 1, 3, 4, 6, 8, 10, 11, 14, 18, and 26.

c. The presidents could have said that they were the *oldest* president (or the same age in years as the oldest) at the time they took office are 1, 2, 7, 9, and 40.

d. Answers will vary.

TECHNOLOGY EXERCISES

53. a

Category of Car	Frequency	Relative Frequency	Cumulative Frequency
American cars	30	37.5%	30
Japanese cars	25	31.25%	55
English cars	5	6.25%	60
Other European cars	12	15%	72
Motorcycles	8	10%	80
Total	80	100%	80

b. 80

c. 100%

UNIT 5D

QUICK QUIZ

1. **c.** The fact that the red segment starts higher up in the second bar indicates the sum of the values represented by the segments below it are greater than those in the first bar.

2. **b.** The height of the Tuberculosis wedge at 1950 is about 20 years.

3. **a.** The color of Oregon corresponds to the 2000 – 2999 gallons of oil equivalent.

4. **c.** Iowa is contained between contour lines that are marked 30°F and 40°F, so the temperature there is between these values.

55.

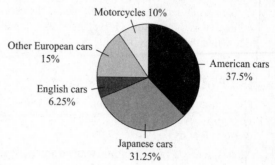

57.

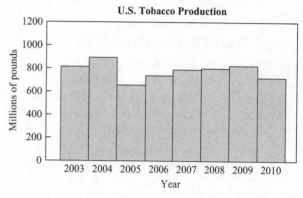

U.S. Tobacco Production

5. **b.** The tan regions lie between contour lines marked by 40°F and 50°F, and thus the temperature there is between those values.

6. **c.** When contour lines are close together, the data they represent change more rapidly compared to areas where the lines are far apart.

7. **a.** The heights of the points on the line chart increase from 2008 to 2012.

8. **b.** Because the scale on the *y*-axis is exaggerated, the changes appear to be larger than they really were.

9. **c.** Each tick mark represents another power of ten.

10. **b.** For all of the years shown in the figure, there is a positive percent change, which means the cost of college was increasing; it increases at a smaller rate when the graph is decreasing.

DOES IT MAKE SENSE?

9. Does not make sense. Using three-dimensional bars for two-dimensional data is a cosmetic change; it adds no information.

11. Makes sense. Experimenting with different scales on the axes of graphs allows you to highlight certain aspects of the data.

BASIC SKILLS AND CONCEPTS

13. a. China, India and Russia all had to import grain because their net grain production is negative.

    b. China, India, and Russia are all expected to need grain imports because their projected 2030 net grain production is negative.

    c. The graph shows that both China and India will require significant grain imports, and thus (based only on the information shown in the graph), it's likely world grain production will need to increase.

15. a. Unemployment decreases with level of education.

    b. The ratio appears the same for the entire graph. Using 2009: $12\% \div 4\% = 3$, so more than 3 times as likely.

17. a. No, the scale on the vertical axis does not start at zero.

    b. Girls would score higher than boys.

19. a. Tuition and fees vary the most; they are almost ten times greater at private 4-year colleges than at public 2-year commuter colleges.

    b. Books and supplies vary the least, probably because students at all types of colleges need the same number of books and other supplies on average, and these don't vary much in cost.

    c. Transportation, because commuters must spend more to get to and from school, while students at private colleges probably spend more time on campus.

21. a. About 14% of the budget went to net interest in 1990; it dropped to about 8% in 2012.

    b. In 1970, about 41% of the budget went to defense. The percentage dropped to about 21% in 2012.

c. About 46% of the budget went to payments to individuals in 1980; it increased to about 53% in 2012.

23. a. Since points A and B correspond to summits, you would walk downhill from A to C.

    b. Since there are more contour lines between B and D, your elevation would change more between those two points.

    c. Since both points lie in the same contour, there would be no change in elevation if you walk directly from E to F.

    d. Since you return to your original location, there would be no net change in elevation.

25. a. In 2010, the over-65-year-old group constituted 12% of the population. By 2050, this figure is projected to increase to about 20%.

    b. The percentage of 45 to 54-year-olds was about 15% in 2010, and is projected to decrease to about 10% in 2050.

    c. The under-25-year-old group decreases in size between 2010 and 2050 since the percentages of each subgroup either decreases or remains constant during that time interval.

    d. The 45 to 54-year-old group was the largest percentage of the population in 2010. (In that year, its bar is higher than all others)

    e. The population is aging as evidenced by the increase in the over-65 category, and the decrease in the under-5 category.

27. The area of the screen on the 2008 TV is larger than the area of the 1980 TV by a factor of about 25, and the volume of the 2008 TV is larger than the volume of the 1980 TV by a factor of about 125, and yet the actual number of homes with cable in 2008 is larger than the 1980 level by a factor of over 5 ($98/18 = 5.44$).

29. The zero point for earnings is not on the graph, and thus it appears that women earn about 25% of what men earn. A fairer graph (not shown) would start from zero.

31. Ordinary vertical scale

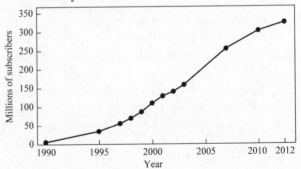

31. (continued)

Exponential vertical scale

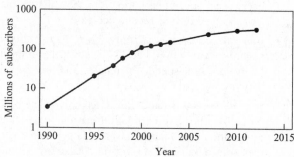

Answers will vary on which scale is more useful as neither scale hides details of the data, although it may be claimed that the ordinary vertical scale may make it appear that the number of subscribers has been increasing linearly.

33. Most of the growth occurred after 1950.

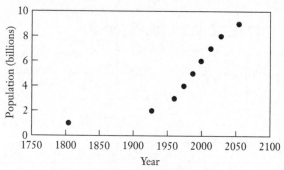

FURTHER APPLICATIONS

35. a. The color of the country indicates the continent in which it is located. Benin is in Africa.

b. Approximately 5 million people in South Africa live with HIV. The per capita income is about $9000.

c. Approximately 1 million people in India live with HIV. The per capita income is about $3000.

d. With some exceptions, HIV incidence decreases with the wealth of the country.

e. South Africa

UNIT 5E

QUICK QUIZ

1. **c.** When one variable is correlated with another, an increase in one goes with either increasing values of the other, or decreasing values of the other.

2. **b.** The dot representing Russia has a horizontal coordinate of about 63 years.

37. A simple bar graph works well for this data. The median age is increasing.

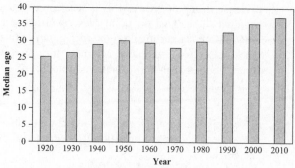

39. A simple multiple bar graph or a line graph would work well for this data. The number of newspapers began decreasing around 1990 while circulation has been decreasing since 1920.

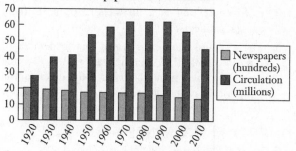

3. **a.** When points on a scatter diagram all lie near a line, there is a strong correlation between the variables represented. We call it a positive correlation when an increase in one variable goes along with an increase in another (and this is what produces a positive slope).

4. **b.** A negative correlation exists between two variables when an increase in one goes with a decrease in the other, and this is what produces a negative slope. That the points described fall in a broad swath, rather than a tight configuration that more closely resembles a line, means the correlation between the variables is weak.

5. **a.** Correlation is a necessary condition for causation.

6. **c.** Exercise tends to decrease body fat, so there would be a negative correlation between body fat and BMI.

7. **a.** When establishing causation, look for evidence that larger amounts of the suspected cause produce larger amounts of the effect.

8. **a.** Extrapolating the upward trend in the data, $2000 is the most likely value.

9. **c.** Since texting is a distraction, it most likely causes accidents.

10. **c.** The courts have proclaimed that a member of the jury should find the defendant guilty when the jury member is firmly convinced the defendant is guilty of the crime charged.

DOES IT MAKE SENSE?

7. Makes sense. A strong negative correlation means that a decrease in the price goes along with an increase in the number of tickets sold.

9. Does not make sense. Despite the fact that A and B are strongly correlated, we cannot be sure that A causes B (there may be an underlying effect that causes both A and B).

11. Makes sense. If it were true that an increase in E caused a decrease in F, then a negative correlation would exist.

BASIC SKILLS AND CONCEPTS

13. We seek a correlation between a car's *weight* and its *city gas mileage*.

    a. The diagram shows a moderately strong negative correlation.

    b. Heavier cars tend to get lower gas mileage.

15. We seek a correlation between *Percent of AGI* and *Average AGI* for 11 states.

    a. At first glance, there's no correlation (which is a reasonable answer), though one could also argue for a very weak positive correlation based on the slight increasing trend of the data points.

    b. Higher AGI may imply slightly higher charitable giving as a percentage of AGI.

17. Use degrees of latitude and degrees of temperature as units. There is a strong negative correlation, because as you move north from the equator, the temperatures in June tend to decrease.

19. Use years and hours as units. There is probably a negative correlation because older people most likely do not spend as much time on social networking sites.

21. Use years and square miles as units. There is probably a positive correlation because states that joined the union later are usually larger than those that joined earlier. (Compare Alaska to Rhode Island, for example.)

23. Use children per woman and years as units. There is a strong negative correlation, because an increase in the life expectancy of a country goes along with a decrease in the number of children born.

FURTHER APPLICATIONS

25. a.

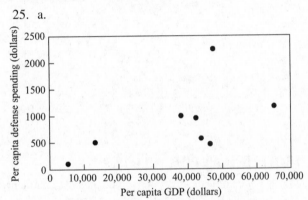

b. There is a strong positive correlation.

c. In those countries where per capita GDP is high, there's more money available to spend on various government programs, and thus the per capita expenditure on defense increases.

27. a.

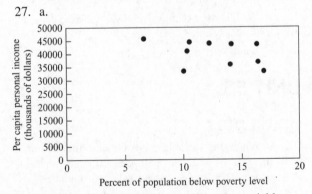

b. There is no correlation between the variables.

NIT 5E: CORRELATION AND CAUSALITY 67

27. (continued)

c. Per capita income will be skewed high (in the direction of wealthier residents of a state), so there will not be correlation with the percent of the population below the poverty level.

29. a.

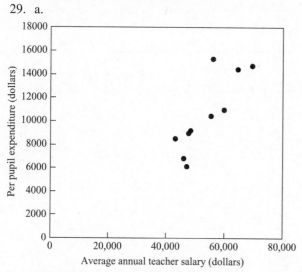

b. There is a strong positive correlation.

c. The per pupil expenditure almost certainly includes teacher salaries (that is, salaries are part of the cost of educating the students), and thus in those districts where the teacher salaries are high, the per pupil expenditure is also high.

31. a. More represented since you senator represents fewer people overall.

b. Representation decreases since the number of people a senator represents increases.

c. Federal aid increases with greater representation.

d. There is a positive correlation between representation and federal funding: As representation increases, federal funding increases.

33. There is a positive correlation between the number of miles of freeway in Los Angeles and the amount of traffic congestion. There is likely a common underlying cause here: an increase in population.

35. Sales of ice cream are positively correlated with sales of swimming suits. The common underlying cause is most likely a change in season.

37. The number of ministers and priests is positively correlated with movie ticket sales. There is a common underlying cause: an increase in population.

39. The causes of cancer are often random. A cancer cell is produced when the growth control mechanisms of a normal cell are altered by a random mutation. Smoking may increase the chance of such a mutation occurring, but the mutation will not occur in all individuals. Therefore, by chance, some smokers escape cancer.

41. You cannot conclude that the high-voltage line is the cause of cancer because a mechanism for the cause must be established. There may be an underlying cause.

opyright © 2015 Pearson Education, Inc.

UNIT 6A

QUICK QUIZ

1. **b.** The mean of a data set is the sum of all the data values divided by the number of data points.

2. **b.** The median of a data set is the middle value when the data are placed in numerical order.

3. **c.** Outliers are data values that are much smaller or larger than the other values.

4. **a.** Outliers that are larger than the rest of the data set tend to pull the mean to the right of the median, and this is the case for all possible data sets that satisfy the conditions given.

5. **c.** Because the mean is significantly higher than the median, the distribution is skewed to the right, which implies some students drink considerably more than 12 sodas per week. Specifically, it can be shown that at least one student must drink more than 16 sodas per week.

6. **c.** The high salaries of a few superstars pull the tail of this distribution to the right.

7. **a.** High outliers tend to pull the mean to the right of the median, and thus you would hope to get a salary near the mean. There are data sets with high outliers that can be constructed where the median is higher than the mean, but these are rare.

8. **a.** In general, data sets with a narrow central peak typically have less variation than those with broad, spread out values (the presence of outliers can confound the issue).

9. **c.** Driving in rush hour can go quickly on some days, but is agonizingly slow on others, so the time it takes to get downtown has high variation.

10. **b.** The mayor would like to see a high median and low variation because that means there's a tight, central peak in the distribution that corresponds to support for her.

DOES IT MAKE SENSE?

7. Makes sense. The two highest grades may be large enough to balance the remaining seven lower grades so that the third highest is the mean.

9. Makes sense. When very high outliers are present, they tend to pull the mean to the right of the median (right-skewed).

11. Does not make sense. When the mean, median, and mode are all the same, it's a sign that you have a symmetric distribution of data.

BASIC SKILLS AND CONCEPTS

13. The mean and median are 53.3 and 52.0, respectively. Because there is an even number of data points, the median is calculated by taking the mean of the middle two values (add 51 and 53, and divide by 2). Since no value occurs more than once, there is no mode.

15. The mean and median and mode are 0.188 and 0.165, respectively. Because there is an even number of data points, the median is calculated by taking the mean of the middle two values (add 0.16 and 0.17, and divide by 2). The mode is 0.16.

17. The mean and median are 58.3 sec and 55.5 sec, respectively. Because there is an even number of data points, the median is calculated by taking the mean of the middle two values (add 53 and 58, and divide by 2). The mode is 49 sec.

19. The mean and median are 0.81237 and 0.8161, respectively. The outlier is 0.7901, because it varies from the others by a couple hundredths of a pound, while the rest vary from each other only by a few thousandths of a pound. Without the outlier, the mean is 0.81608, and the median is 0.8163.

21. The median would be a better representation of the average. A small number of households with high earnings skew the distribution to the right, affecting the mean.

23. This distribution is probably skewed to the right by a small percentage of people who change jobs many times. Thus the median would do a better job of representing the average, as the outliers affect the mean.

25. This is a classic example of a symmetric distribution, and thus the mean and the median are almost identical (in a perfectly symmetric distribution, they are identical) – either the mean or the median could be used.

27. a. There is one peak on the right due to all the students who received A's.

    b.
    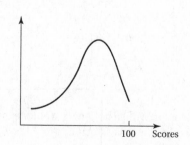

27. (continued)

 c. The distribution is left-skewed because the scores tail off to the left.

 d. The variation is large.

29. a. There is likely one peak since most cities would have similar amounts of annual rainfall.

 b.

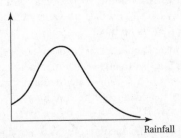

Rainfall

 c. The distribution would be right-skewed because a few cities with higher rainfall would vary from the average more than cities with lower rainfall.

 d. The variation would be large since some cities would have rainfall amounts much lower or much higher than the mean rainfall amount.

31. a. There would be one peak since sales would be high in the winter months, with very low sales (or even no sales) in the summer.

 b.

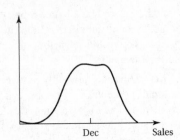

Dec   Sales

 c. This is a symmetric distribution.

 d. The variation is moderate because most of the sales occur in the winter.

33. a. There would be one peak in the distribution, corresponding to the average price of dog food.

 b.

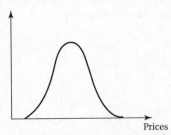

Prices

 c. More likely than not, there would be no extreme outliers in this data set, and it would be symmetric.

d. There would be moderate variation in the data set because we don't expect outliers. (The dog food may not sell well next to 19 other brands if its price is radically different. Of course this begs the question, "But what if it's much cheaper than the other brands?" That's not too likely to happen for a commodity like dog food, because once one company begins selling its dog food at a low price, the others would be forced to lower their prices to compete).

35. The distribution would be left-skewed since the mean is lower than the median. The variation would be high given the low score of 22 and the high score of 100 are both far from the average score.

FURTHER APPLICATIONS

37. The distribution has two peaks (bimodal), with no symmetry, and large variation. For tourists who want to watch the geyser erupt, it means they may have to wait around for a while (most likely between 25 and 40 minutes, though perhaps as long as 50 to 90 minutes, depending on when they arrive at the site).

Times Between Eruptions of Old Faithful

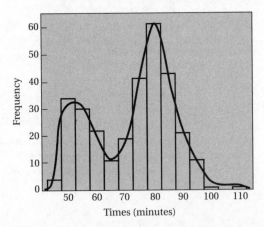

39. This distribution has one peak, is symmetric, and has moderate variation. The graph says that weights of rugby players are usually near the mean weight, but that there are some light and some heavy players.

39. (continued)

Weight of Rugby Players

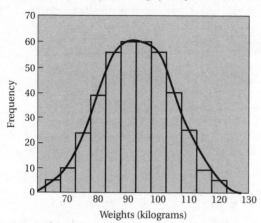

UNIT 6B

QUICK QUIZ

1. **a**. The range is defined as the high value – low value.

2. **c**. The five-number summary includes the low value, the first quartile, the median, the third quartile, and the high value. Unless the mean happens to be the same as the median, it would not be part of the five-number summary.

3. **a**. Roughly half of any data set is contained between the lower and upper quartiles ("roughly" because with an odd number of data values, one cannot break a data set into equal parts).

4. **b**. Consider the set {0, 0, 0, 0, 0, 0, 0, 0, 0, 100}. Its mean is 10, which is larger than the upper quartile of 0.

5. **b**. You need the high and low values to compute the range, and you need all of the values to compute the standard deviation, so the only thing you can compute is a single deviation.

6. **b**. The standard deviation is defined in such a way that it can be interpreted as the average distance of a random value from the mean.

7. **c**. The standard deviation is defined as the square root of the variance, and thus is always non-negative.

8. **a**. Using the range rule of thumb, we see that the range is about four times as large as the standard deviation. (The worst-case scenario is when the range and standard deviation are both zero).

9. **b**. Newborn infants and first grade boys have similar heights, whereas there is considerable variation in the heights of all elementary school children.

41. a. Sketches will vary. We cannot be sure of the exact shape of the distribution with the given parameters, as we don't have all the raw data. However, because the mean is larger than the median, and because the data set has large outliers, it would likely be right-skewed, with a single peak at the mode.

    b. About 150 families (50%) earned less than $45,000 because that value is the median income.

    c. We don't have enough information to be certain how many families earned more than $55,000, but it is likely a little less than half, simply because $55,000 is greater than the median.

43. $(3 \times 3.7 + 3 \times 3.4 + 4 \times 2.7 + 3 \times 3 + 1 \times 4) \div (3 + 3 + 4 + 3 + 1) = 45.1 \div 14 = 3.22$

10. **c**. Because the standard deviation for Garcia is the largest, the data are more spread out, so Professor Garcia must have had very high grades from more students than the other two.

DOES IT MAKE SENSE?

7. Does not make sense. The range depends upon only the low and high values, not on the middle values.

9. Makes sense. Consider the case where the 15 highest scores were 80, the next score was 68, and the lowest 14 scores were 40. There are numerous such cases that satisfy the conditions given.

11. Makes sense. The standard deviation describes the spread of the data, and one would certainly expect the heights of 5-year old children to have less variation than the heights of children aged 3 to 15.

BASIC SKILLS AND CONCEPTS

13. The mean waiting time at Big Bank is (4.1 + 5.2 + 5.6 + 6.2 + 6.7 + 7.2 + 7.7 + 7.7 + 8.5 + 9.3 + 11.0) ÷ 11 = 79.2 ÷ 11 = 7.2. Because there are 11 values, the median is the sixth value = 7.2.

15. a. For the East Coast, the mean, median, and range are 157.7, 131.25, and 209.4, respectively. For the Midwest, the mean, median, and range are 115.8, 94.9, and 140.8, respectively.

    b. For the East Coast, the five-number summary is (104.8, 123.4, 131.25, 141.1, 314.2). For the Midwest, it is (87.4, 92.9, 94.9, 96.5, 228.2).

15. (continued)

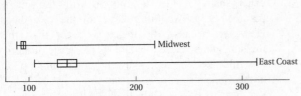

c. The standard deviation for the East Coast is 77.67. For the Midwest, it is 55.16.

d. For the East Coast, the range rule of thumb estimates the standard deviation as 209.4 ÷ 4 = 52.35, which is a far cry from the actual value of 77.67. This is due, in part, to the outlier of New York City (314.2). For the Midwest, the range rule of thumb estimates the standard deviation as 140.8 ÷ 4 = 35.2, which is also well off the mark.

e. The cost of living index is smaller, on average, for the Midwest cities shown. The variation is also lower (due in part to the outlier of New York City).

17. a. For the no treatment group, the mean, median, and range are 0.184, 0.16, and 0.35, respectively. For the treatment group, the mean, median, and range are 1.334, 1.07, and 2.11, respectively.

b. The five-number summary for the no treatment group is (0.02, 0.085, 0.16, 0.295, 0.37). It is (0.27, 0.595, 1.07, 2.205, 2.38) for the treatment group.

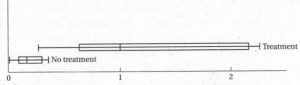

c. The standard deviation for the no treatment group is 0.127; for the treatment group, it is 0.859.

d. The range rule of thumb estimates the standard deviation for the no treatment group as 0.35 ÷ 4 = 0.0875, which underestimates the actual value of 0.127. For the treatment group, the range rule of thumb estimates the standard deviation as 2.11 ÷ 4 = 0.5275, which is underestimates the actual value of .859.

e. The average weight and standard deviation of trees in the treatment group are higher that those of the no treatment group.

FURTHER APPLICATIONS

19. a.

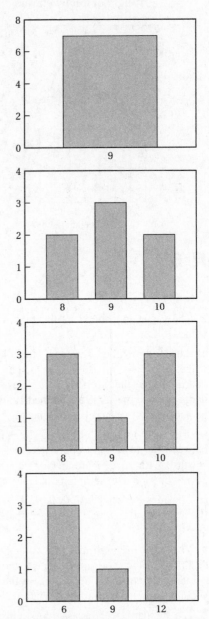

b. The five-number summaries for each of the sets shown are (in order): (9, 9, 9, 9, 9); (8, 8, 9, 10, 10); (8, 8, 9, 10, 10); and (6, 6, 9, 12, 12). The boxplots are not shown.

c. The standard deviations for the sets are (in order): 0.000, 0.816, 1.000, and 3.000.

d. Looking at just the answers to part c, we can tell that the variation gets larger from set to set.

21. The first pizza shop has a slightly larger mean delivery time, but much less variation in delivery time when compared to the second shop. If you want a reliable delivery time, choose the first shop. If you don't care when the pizza arrives, choose the shop that offers cheaper pizza (you like them equally well, so you may as well save some money).

23. A lower standard deviation means more certainty in the return, and a lower risk.

25. Batting averages are less varied today than they were in the past. Because the mean is unchanged, batting averages above .350 are less common today.

27. a. For Healing hospital, the mean and median are 8.26 lb and 8.6 lb, respectively. For Healthy hospital, the mean and median are 7.27 lb and 7.2 lb, respectively.

    b. For Healing hospital, the standard deviation is 0.88 lb. For Healthy hospital, the standard deviation is 0.88 lb.

    c.

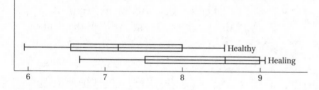

29. a. For Supplier A, the mean and standard deviation are 16.32 mm and 0.057 mm, respectively. For Supplier B, the mean and standard deviation are 16.33 mm and 0.098 mm, respectively.

    b.

    c. Supplier A: $6 \div 7 \approx 86\%$

    Supplier A: $4 \div 7 \approx 57\%$

UNIT 6C

QUICK QUIZ

1. **b**. A normal distribution is symmetric with one peak, and this produces a bell-shaped curve.

2. **a**. The mean, median and mode are equal in a normal distribution.

3. **a**. Data values farther from the mean correspond to lower frequencies than those close to the mean, due to the bell-shaped distribution.

4. **c**. Since most of the workers earn minimum wage, the mode of the wage distribution is on the left, and the distribution is right-skewed.

5. **a**. Roughly 68% of data values fall within one standard deviation of the mean, and 68% is about 2/3.

6. **c**. In a normally distributed data set, 99.7% of all data fall within 3 standard deviations of the mean.

7. **a**. Note that 43 mpg is 1 standard deviation above the mean. Since 68% of the data lies within one standard deviation of the mean, the remaining 32% is farther than 1 standard deviation from the mean. The normal distribution is symmetric, so

half of 32%, or 16%, lies above 1 standard deviation.

8. **c**. The z-score for 84 is $(84 - 75)/6 = 1.5$.

9. **c**. If your friend said his IQ was in the 75th percentile, it would mean his IQ is larger than 75% of the population. It is impossible to have an IQ that is larger than 102% of the population.

10. **b**. Table 6.3 shows that the percentile corresponding with a standard score of –0.60 is 27.43%.

DOES IT MAKE SENSE?

5. Makes sense. Physical characteristics are often normally distributed, and college basketball players are tall, though not all of them, so it makes sense that the mean is 6'3", with a standard deviation of 3" (this would put approximately 95% of the players within 5'9" to 6'9").

7. Does not make sense. With a mean of 6.8 pounds, and a standard deviation of 7 pounds, you would expect that a small percentage of babies would be born 2 standard deviations above the mean. But this would imply that some babies weigh more than 20.8 pounds, which is much too heavy for a newborn.

9. Makes sense. A standard score of 2 or more corresponds to percentiles of 97.72% and higher, so while this could certainly happen, the teacher is giving out As (on the final exam) to only 2.28% of the class

BASIC SKILLS AND CONCEPTS

11. Diagrams **a** and **c** are normal, and diagram **c** has a larger standard deviation because its values are more spread out.

13. Most trains would probably leave near to their scheduled departure times, but a few could be considerably delayed, which would result in a right-skewed (and non-normal) distribution.

15. Random factors would be responsible for the variation seen in the distances from the bull's-eye, so this distribution would be normal.

17. Difficult exams would have scores that are right-skewed, so the scores would not be normally distributed.

19. a. Half of any normally distributed data set lies below the mean, so 50% of the scores are less than 100.

    b. Note that 120 is one standard deviation above the mean. Since 68% of the data lies within one standard deviation of the mean (between 80 and 120), half of 68%, or 34%, lies between 100 and 120. Add 34% to 50% (the answer to part a) to find that 84% of the scores are less than 120.

    c. Note that 140 is two standard deviations above the mean. Since 95% of the data lies within two standard deviations above the mean (between 60 and 140), half of 95%, or 47.5%, lies between 100 and 140. Add 47.5% to 50% (see part a) to find that 97.5% of the scores are below 140.

    d. Note that 60 is two standard deviations below the mean. Since 95% of the data lies within two standard deviations of the mean (between 60 and 140), 5% of the data lies outside this range. Because the normal distribution is symmetric, half of 5%, or 2.5%, lies in the lower tail below 60.

    e. A sketch of the normal distribution would show that the amount of data lying above 60 is equivalent to the amount of data lying below 140, so the answer is 97.5% (see part c).

    f. Note that 160 is three standard deviations above the mean. Since 99.7% of the data lies within three standard deviations of the mean (between 40 and 160), 0.3% of the data lies outside this range. Because the normal distribution is symmetric, half of 0.3%, or 0.15%, lies in the upper tail above 160.

    g. A sketch of the normal distribution shows that the percent of data lying above 80 is equivalent to the percent lying below 120, so the answer is 84% (see part b).

    h. The percent of data between 60 and 140 is 95%, because these are each two standard deviations from the mean.

21. A score of 59 is one standard deviation below the mean. Since 68% of normally distributed data lies within one standard deviation of the mean, half of 68%, or 34%, lies between 59 and 67 due to symmetry. The percent of data lying above 67 is 50%, so we add these two results to find that 84% of the scores lie above 59.

23. Subtract two standard deviations from the mean to get the cutoff score of $67 - 16 = 51$. Since 95% of the data lies within two standard deviations of the mean, 5% lies outside this range. Half of 5%, or 2.5%, will be below 51, and will fail the exam.

25. The standard score for 67 is $z = (67 - 67)/8 = 0$.

27. The standard score for 55 is $z = (55 - 67)/8 = -1.5$

29. a. $z = 1$. The corresponding percentile is 84.13%.

    b. $z = 0.5$, and the corresponding percentile is 69.15%.

    c. $z = -1.5$, and the corresponding percentile is 6.68%.

31. a. The 20th percentile corresponds to a standard score of about $z = -0.83$, which means the data value is about 0.83 standard deviations below the mean.

    b. The 80th percentile corresponds to a standard score of about $z = 0.84$, which means the data value is 0.84 standard deviations above the mean.

    c. The 63rd percentile corresponds to a standard score of about $z = 0.33$, which means the data value is 0.33 standard deviations above the mean.

FURTHER APPLICATIONS

33. About 68% of births occur within 15 days of the mean, because this range is one standard deviation from the mean.

35. Since 68% of births occur within 15 days of the due date, 32% occur outside this range, and half of 32%, or 16%, occur more than 15 days after the due date.

37. a. The standard score is $z = (65 - 63.6)/2.5 = 0.56$, which corresponds to an approximate percentile of 71%.

    b. $z = (62 - 63.6)/2.5 = -0.64$, which corresponds to an approximate percentile of 26%.

    c. $z = (60.5 - 63.6)/2.5 = -1.24$, which corresponds to an approximate percentile of 11%.

    d. $z = (65.7 - 63.6)/2.5 = 0.84$, which corresponds to an approximate percentile of 80%.

39. It is not likely because heights of eighth-graders would be normally distributed, and a standard deviation of 40 inches would imply some of the eighth graders (about 16%) are taller than 95 inches, which is preposterous.

41. The 90th percentile corresponds to a z-score of about 1.3, which means a data value is 1.3 standard deviations above the mean. For the verbal portion of the GRE, this translates into a score of $462 + 1.3(119) = 617$. For the quantitative portion, it translates into a score of $584 + 1.3(151) = 780$.

43. The standard score for 650 is $z = (650 - 584)/151 = 0.44$, which corresponds to about the 67th percentile. This is the percent of students who score below 650, so the percent that score above 650 is 33%.

45. The z-score for an 800 on the quantitative exam is $z = (800 - 584)/151 = 1.43$, or about 1.4. This corresponds to the 91.92 percentile, so about 92% score below 800, which means 8% score 800.

UNIT 6D

QUICK QUIZ

1. **c.** A divorce rate of 60% is statistically significant because such large variation from the national rate of 30% is not likely to occur by chance alone.

2. **b.** In order to determine whether a result is statistically significant, we must be able to compare it to a known quantity (such as the effect on a control group in this case).

3. **a.** Statistical significance at the 0.01 level means an observed difference between the effectiveness of the remedy and the placebo would occur by chance in fewer than 1 out of 100 trials of the experiment.

4. **c.** If the statistical significance is at the 0.05 level, we would expect that 1 in 20 experiments would show the pill working better than the placebo by chance alone.

5. **b.** The confidence interval is found by subtracting and adding 3% to the results of the poll (65%).

47. For the verbal exam, $z = (450 - 462)/119 = -0.10$, which corresponds to an approximate percentile of 46%. Thus about 46% score below 450. For the quantitative exam, $z = (450 - 584)/151 = -0.89$, which corresponds to an approximate percentile of 19%. Thus about 19% score below 450.

TECHNOLOGY EXERCISES

53. a. $z = (72 - 69.7)/2.7 = 0.85$, which, from table 6.3, corresponds to an approximate percentile of 80%.

    b. $z = (65 - 69.7)/2.7 = -1.74$; NORMDIST(64,69.7,2.7,TRUE) gives the percentile as 4.0865%, which corresponds to an approximate percentile of 4%.

    c. $6'4" = 76"$; NORMDIST(76,69.7,2.7,TRUE), gives the percentile as 99.0185%, so approximately 1% of American adult males are taller than $6'4"$.

    d. $5'4" = 64"$; NORMDIST(64,69.7,2.7,TRUE), gives the percentile as 1.7381%, so approximately 1.7% of American adult males are shorter than $5'4"$.

    e. $6'7" = 79"$; NORMDIST(79,69.7,2.7,TRUE), gives the percentile as 99.97139%, so approximately (100% - 99.97139%) × 115,000,000 = 33,000 of American adult males are as tall as Kobe Bryant.

6. **c.** The central limit theorem tells us that the true proportion is within the margin of error 95% of the time.

7. **c.** The margin of error is approximately $\dfrac{1}{\sqrt{n}}$. Set this equal to 4% (0.04), and solve for n to find that the sample size was 625. If we quadruple the sample size to 2500, the margin of error will be approximately $\dfrac{1}{\sqrt{2500}} = 0.02$, or 2%.

8. **a.** The null hypothesis is often taken to be the assumption that there is no difference in the things being compared.

9. **c.** If you cannot reject the null hypothesis, the only thing you can be sure of is that you don't have evidence to support the alternative hypothesis.

10. **b**. If the difference in gas prices can be explained by chance in only 1 out of 100 experiments, you've found a result that is statistically significant at the 0.01 level.

DOES IT MAKE SENSE?

9. Makes sense. If the number of people cured by the new drug is not significantly larger than the number cured by the old drug, the difference could be explained by chance alone, which means the results are not statistically significant.

11. Does not make sense. The margin of error is based upon the number of people surveyed (the sample size), and thus both Agency A and B should have the same margin of error.

13. Makes sense. The alternative hypothesis is the claim that is accepted if the null hypothesis is rejected.

BASIC SKILLS AND CONCEPTS

15. Most of the time you would expect around 10 sixes in 60 rolls of a standard die. It would be very rare to get only 2 sixes, and thus this is statistically significant.

17. Note that 11/200 is about 5% – this is exactly what you would expect by chance alone from an airline that has 95% of its flights on time.

19. There aren't many winners in the Power Ball lottery, and yet there are a good number of 7-11 stores. It would be very unlikely that ten winners in a row purchased their tickets at the same store, so this is statistically significant.

21. The result is significant at far less than the 0.01 level (which means it is significant at both the 0.01 and 0.05 level), because the probability of finding the stated results is less than 1 in 1 million by chance alone. This gives us good evidence to support the alternative hypothesis that the accepted value for temperature is wrong.

23. The results of the study are significant at neither the 0.05 level nor the 0.01 level, and thus the improvement is not statistically significant.

25. The margin of error is $1/\sqrt{1012} = 0.031$, or 3.1%, and the 95% confidence interval is 28.% to 35.1%. This means we can be 95% confident that the actual percentage is between 28.9% and 35.1%.

27. The margin of error is $1/\sqrt{1504} = 0.026$, or 2.6%, and the 95% confidence interval is 45.4% to 50.6%. This means we can be 95% confident that the actual percentage is between 45.4% and 50.6%.

29. The margin of error is $1/\sqrt{1003} = 0.032$, or 3.2%, and the 95% confidence interval is 44.8% to 51.2%. This means we can be 95% confident that the actual percentage is between and 44.8% and 51.2%.

31. The margin of error is $1/\sqrt{1020} = 0.031$, or 3.1%, and the 95% confidence interval is 70.9% to 77.1%. This means we can be 95% confident that the actual percentage is between and 70.9% and 77.1%.

33. a. Null hypothesis: six-year graduation rate = 42%. Alternative hypothesis: six-year graduation rate > 42%.

    b. Rejecting the null hypothesis means there is evidence that the graduation rate exceeds 42%. Failing to reject the null hypothesis means there is insufficient evidence to conclude that the graduation rate exceeds 42%.

35. a. Null hypothesis: mean teacher salary = $47,750. Alternative hypothesis: mean teacher salary > $47,750.

    b. Rejecting the null hypothesis means there is evidence that the mean teacher salary in the state exceeds $47,750. Failing to reject the null hypothesis means there is insufficient evidence to conclude that the mean teacher salary exceeds $47,750.

37. a. Null hypothesis: percentage of underrepresented students = 20%. Alternative hypothesis: percentage of underrepresented students < 20%.

    b. Rejecting the null hypothesis means there is evidence that the percentage of underrepresented students is less than 20%. Failing to reject the null hypothesis means there is insufficient evidence to conclude that the percentage of underrepresented students is less than 20%.

39. Null hypothesis: mean annual mileage of cars in the fleet = 11,725 miles. Alternative hypothesis: mean annual mileage of cars in the fleet > 11,725. The result is significant at the 0.01 level, and provides good evidence for rejecting the null hypothesis.

41. Null hypothesis: mean stay = 2.1 days. Alternative hypothesis: mean stay > 2.1 days. The result is not significant at the 0.05 level, and there are no grounds for rejecting the null hypothesis.

43. Null hypothesis: mean income = $50,000. Alternative hypothesis: mean income > $50,000. The result is significant at the 0.01 level, and provides good evidence for rejecting the null hypothesis.

FURTHER APPLICATIONS

45. The margin of error is $1/\sqrt{5000} = 0.014$, or 1.4%, and the 95% confidence interval is 11.5% to 14.3%. This means we can be 95% confident that the actual percentage of viewers who watched *CSI* was between 11.5% and 14.3%.

47. The margin of error is approximately $1/\sqrt{n}$. If we want to decrease the margin of error by a factor of 2, we must increase the sample size by a factor of 4, because $\dfrac{1}{\sqrt{4n}} = \dfrac{1}{2\sqrt{n}} = \dfrac{1}{2} \cdot \dfrac{1}{\sqrt{n}}$.

49. The margin of error, 3%, is consistent with the sample size because $1/\sqrt{1019} = 0.03 = 3\%$.

UNIT 7A

QUICK QUIZ

1. **b**. HHT is a different outcome than HTH due to the order in which the heads appear, but both are the same event because each has two heads.

2. **b**. This is a probability based on observations about the number of free throws Shawna attempted and made, and so it is empirical.

3. **a**. This is a theoretical probability because we are assuming each outcome is equally likely, and no experiment is being carried out to determine the probability.

4. **c**. If the probability of an event is $P(A)$, the probability it does not occur is $1 - P(A)$.

5. **c**. The most likely outcome when a coin is tossed several times is that half of the time heads will appear.

6. **b**. Table 7.3 shows that the probability Mackenzie wins is 5/36, whereas the probability that Serena wins is 4/36.

7. **c**. There are six possible outcomes for each of the four dice, and the multiplication principle says we multiply these outcomes together to get the total number of outcomes when all four dice are rolled.

8. **b**. The lowest sum that could occur when four dice are rolled is 4, and the highest is 24, and every sum in between is possible.

9. **a**. The sum of all the probabilities for a probability distribution is always 1.

10. **a**. If Triple Treat's probability of winning is 1/4, then the probability he loses is 3/4. This implies the odds that he wins is $(1/4) \div (3/4)$, or 1 to 3, which, in turn, means the odds against are 3 to 1.

DOES IT MAKE SENSE?

7. Makes sense. The outcomes are HTTT, THTT, TTHT, TTTH.

9. Makes sense. This is a subjective probability, but it's as good as any other guess one might make.

11. Does not make sense. The sum of $P(A)$ and $P(\text{not } A)$ is always 1.

BASIC SKILLS AND CONCEPTS

13. There are $15 \times 3 = 45$ choices.

15. There are $3 \times 6 \times 5 = 90$ three-course meals.

17. The possible outcomes for the two days are (SS, SC, CS, CC). If only concerned with the number of sunny days, the outcomes are (0,1,2).

19. The possible outcomes for the two tosses are (HH, HT, TH, TT). The probability is 3/4. We are assuming equally likely outcomes.

21. There are 4 queens in a standard deck, so the probability is 4/52 = 1/13. We are assuming that each card is equally likely to be chosen.

23. There are 4 possible outcomes with a 2-child family (BB, BG, GB, GG), and 2 of these have one boy and one girl, so the probability is 2/4 = 1/2. This assumes each outcome is equally likely to occur.

25. There are 12 months in the year, so the probability is 1/12, assuming births are uniformly distributed across the months of the year. If using the number of days in the months, there are 31 total days in January, so the probability is 31/365, assuming births are uniformly distributed across the days of the year.

27. There are 31 days in October, so the probability is 1/31, assuming births are uniformly distributed across the days of the month.

29. The relative frequency probability is 18/30 = 3/5 = 0.6.

31. The relative frequency probability is 12/100 = 3/25 = 0.12. Since you would expect the probability to be 0.50, you should suspect the coin is unfair.

33. Theoretical

35. The probability is 5/6, because there are 5 outcomes that aren't 2s. It is assumed that each outcome is equally likely.

37. The probability he will make the next free throw is 75% = 0.75, so the probability he will miss is 1 – 0.75 = 0.25, or 25%. The probabilities computed here are based on his past performance.

39.

Result	Probability
0 B	1/8
1 B	3/8
2 B	3/8
3 B	1/8

41.

Sum	Probability
2	1/16
3	2/16
4	3/16
5	4/16
6	3/16
7	2/16
8	1/16

43. The probability of getting a 1 or a 2 is 2/6 = 1/3, and the probability of not getting a 1 or a 2 is 2/3. Thus the odds for the event are (1/3) ÷ (2/3) = 1 to 2, and the odds against are 2 to 1.

45. This problem is similar to Exercise 43, except that we are focusing on getting a 5 or a 6 (instead of a 1 or a 2) – the probabilities and odds remain the same. Thus the odds for the event are 1 to 2, and odds against are 2 to 1.

47. Since the odds on your bet are 3 to 4, you'll win $3 for every $4 you bet. A $20 bet is five $4 bets, so you will gain $15.

FURTHER APPLICATIONS

49. There are 12 royalty cards, so the probability is 12/52 = 3/13. It is assumed that each card is equally likely to be chosen.

51. The probability is 2/52 = 1/26 because there are 2 red aces. We are assuming equally likely outcomes.

53. Assuming that people are equally likely to be born during any hour of the day, the probability is 1/24.

55. There are ten possible outcomes for the last digit in a phone number, and two of these is are 1 or 2, so the probability is 2/10 = 1/5. It is assumed that each digit is equally likely to occur at the end of a phone number.

57. There are 8 possible outcomes with a 3-child family (BBB, BBG, BGB, GBB, GGB, GBG, BGG, GGG), and 1 of these has exactly three girls, so the probability is 1/8. This assumes each outcome is equally likely to occur.

59. There are ten possible outcomes for the last digit in a social security number, so the probability is 1/10, assuming that final digits are uniformly distributed over social security numbers.

61. Answers will vary as this is a subjective probability.

63. The probability of rolling a sum of 7 is 6/36 (see Table 7.3), so the probability of not rolling a sum of 7 is 1 – 6/36 = 30/36 = 5/6. Equally likely outcomes are assumed.

65. The probability of having a 100-year flood is 1/100, so the probability of not having one is 1 – 1/100 = 99/100 = 0.99. This assumes that each year the probability of a 100-year flood is 1/100.

67. There are 6 × 6 = 36 possible outcomes, and 3 of these sum to 10 (see Table 7.3), so the probability is 3/36 = 1/12. It is assumed that each outcome is equally likely to occur.

69. a. {WW, WB, WR, WG, BB, BW, BR, BG, RR, RW, RB, RG, GG, GW, GR, GB}

b. Each outcome is equally likely.

Result	Probability
0 B	9/16
1 B	6/16 = 3/8
2 B	1/16

71. There are 4 × 3 × 5 × 2 × 3 = 360 sets.

73. There are 2 × 2 × 2 × 8 = 64 versions.

75. The probability in 2010 is 40 million ÷ 310 million = 0.13, and the probability in 2050 is 82 million ÷ 439 million = 0.19. Thus the probability is greater in 2050.

77. The odds for event A are (0.99) ÷ (0.01) = 99 to 1, and the odds for event B are (0.96) ÷ (0.04) = 24 to 1. The odds for A are much higher, even though the probabilities for A and B are relatively close to one another.

79. Answers will vary. Theoretical outcomes are:

Result	Probability
0 H	1/8
1 H	3/8
2 H	3/8
3 H	1/8

UNIT 7B

QUICK QUIZ

1. **c.** The probability of getting a double-1 on at least one of the rolls is $1 - (35/36)^2$.

2. **c.** The rule works only when A and B are independent events.

3. **b.** The probability of choosing a pink rose on the second pick is dependent upon what happens on the first pick (because there will be fewer roses in the vases).

4. **c.** Since these are dependent events, the rule $P(A$ and B and $C) = P(A)$ $P(B$ given $A)$ $P(C$ given A and $B)$ must be used.

5. **c.** A person can be born on both a Wednesday and in July, so the events are overlapping.

6. **c.** $P(5$ or 6 or 7 or $8) = P(5) + P(6) + P(7) + P(8) = 4/36 + 5/36 + 6/36 + 5/36 = 20/36 > 0.5$.

7. **a.** Since these events are independent, we have $P(sum\ of\ 3$ and $sum\ of\ 4) = P(sum\ of\ 3) \times P(sum\ of\ 4) = 2/36 \times 3/36$.

8. **b.** The probability of getting at least one success in ten trials of the experiment is $1 - [P(\text{not two heads})]^{10}$, so you need first to compute the probability of not two heads.

9. **c.** This answer shows the correct application of the *at least once* rule.

10. **c.** This answer shows the correct application of the *at least once* rule.

DOES IT MAKE SENSE?

5. Makes sense. You can't get both heads and tails on a single flip of a coin, but you must get one or the other.

7. Does not make sense. The probability of drawing an ace or a spade is 16/52, but the probability of drawing the ace of spades is 1/52.

9. Does not make sense. The chance of getting at least one 5 is $1 - (5/6)^3 \neq 3/6$.

BASIC SKILLS AND CONCEPTS

11. Results will vary. You will not always get at least one head in your two rolls.

13. The events are most likely independent (there may be some situations where one could argue dependence, such as the case where a woman gives birth to quintuplets, and these happen to be the next five births – but we can't solve the problem without assuming independence). Assuming the probability of giving birth to a girl is 1/2, the answer is $(1/2)^4 = 1/16$.

15. The events are independent. $P(2$ and 2 and $6) = P(2) \times P(2) \times P(6) = (1/6)^3 = 1/216$.

17. The events are dependent. $P(R_1$ and $R_2) = P(R_1) \times P(R_2$ given $R_1) = (8/40) \times (7/39) = 7/195$.

19. The events are dependent, and the probability is $(15/30) \times (14/29) \times (13/28) \times (12/27) \times (11/26) \times (10/25) \times (9/24) \approx 0.0032$.

21. The events are non-overlapping. $P(10$ or 11 or $12) = P(10) + P(11) + P(12) = 1/36 + 2/36 + 3/36 = 1/6$.

23. The events are non-overlapping. $P(red\ 6$ or $black\ 8) = P(red\ 6) + P(black\ 8) = 2/52 + 2/52 = 1/13$.

25. The events are independent for all practical purposes, because the batch from which the chips are chosen is large. The probability is $(0.95)^3 \approx 0.857$.

27. $P(\text{at least one } T) = 1 - (1/2)^4 = 15/16$.

29. $P(\text{rain at least once}) = 1 - (0.80)^6 \approx 0.738$.

31. (at least one win) $= 1 - (0.99)^{10} \approx 0.0956$.

FURTHER APPLICATIONS

33. The events are independent. Because the probability of rolling an even result is the same on each throw (1/2), the probability of getting an odd result on all four dice is $(1/2)^4 = 1/16$.

35. The events are overlapping. The probability is $1/6 + 3/6 - 1/6 = 3/6 = 1/2$.

37. The probability of not getting a 6 on a single roll is 5/6, so $P(\text{at least one } 6) = 1 - (5/6)^4 \approx 0.518$.

39. The events are independent. This is similar to Exercise 16, with an answer of $(13/52)^3 = 1/64 = 0.0156$.

41. The events are overlapping. $P(face\ card$ or $diamond) = P(face\ card) + P(diamond) - P(face\ card$ and $diamond) = 12/52 + 13/52 - 3/52 = 22/52 = 11/26$.

43. The events are dependent. $P(Q_1$ and Q_2 and $Q_3) = P(Q_1) \times P(Q_2$ given $Q_1) \times P(Q_3$ given Q_1 and $Q_2) = (4/52) \times (3/51) \times (2/50) = 1/5525 \approx 0.0002$.

45. $P(\text{at least one king}) = 1 - (48/52)^8 \approx 0.473$.

47. The events are independent. The probability of all four tickets being winners is $(1/10)^4 = 1/10,000$.

49. $P(\text{at least one lefty}) = 1 - (0.89)^8 \approx 0.606$.

51. The events are overlapping. $P(W \text{ or } R) = P(W) + P(R) - P(W \text{ and } R) = 140/200 + 120/200 - 80/200 = 180/200 = 9/10 = 0.9$.

53. The events are non-overlapping. The easiest way to solve the problem is to look at the set of possible outcomes for a three child family, (GGG, GGB, GBG, BGG, BBG, BGB, GBB, BBB), and count the number of outcomes with either 1 or 2 boys. The probability is $6/8 = 3/4$. One could also use the formulas for *or* probabilities.

55. $P(\text{at least one flu encounter}) = 1 - (0.96)^{12} \approx 0.387$.

57. a. $P(G \text{ or } P) = P(G) + P(P) - P(G \text{ and } P) = 956/1028 + 450/1028 - 392/1028 \approx 0.986$.

    b. $P(NG \text{ or } NP) = P(NG) + P(NP) - P(NG \text{ and } NP) = 72/1028 + 578/1028 - 14/1028 \approx 0.619$.

59. a. These events are dependent because there are fewer people to choose from on the second and subsequent calls, and this affects the probability that a Republican is chosen.

    b. $P(R_1 \text{ and } R_2) = P(R_1) \times P(R_2 \text{ given } R_1) = (25/45) \times (24/44) \approx 0.303$.

c. If the events are treated as independent, then the pollster is selecting a person from random and not crossing that person off the list of candidates for the second call. The probability of selecting two Republicans is then $P(R_1 \text{ and } R_2) = P(R_1) \times P(R_2) = (25/45) \times (25/45) \approx 0.309$.

d. The probabilities are nearly the same. This is due to the fact that the list of names is large, and removing one person from the list before making the second call doesn't appreciably change the probability that a Republican will be chosen.

61. The probability of winning would be $1 - (35/36)^{25} \approx 0.506$. Since this is larger than 50%, he should come out ahead over time.

63. a. The probability is 1/20.

    b. $P(H_1 \text{ and } H_2) = P(H_1) \times P(H_2) = (1/20) \times (1/20) = 1/400 = 0.0025$.

    c. $P(\text{at least one hurricane}) = 1 - (19/20)^{10} \approx 0.401$.

65. $P(\text{misses } 1^{st} \text{ free throw}) = 1 - 0.7 = 0.3$.

    $P(\text{makes } 1^{st} \text{ and misses } 2^{nd}) = (0.7)(0.3) = 0.21$.

    $P(\text{makes } 1^{st} \text{ and makes } 2^{nd}) = (0.7)(0.7) = 0.49$.

    Making both free throws is most likely.

UNIT 7C

QUICK QUIZ

1. **b**. The law of large numbers says that the proportion of years should be close to the probability of a hurricane on a single year (and $100/1000 = 0.1$).

2. **c**. The expected value is $-\$1 \times 1 + \$75$ million $\times 1/100,000,000 = -\$0.25$.

3. **c**. If you purchase 1/10 of the available number of lottery tickets, there's a 1 in 10 chance that you have the winning ticket. But this means that 90% of the time you have purchased 1 million losing tickets.

4. **c**. The expected value can be understood as the average value of the game over many trials.

5. **b**. The expected value of a $2000 fire insurance policy from the point of view of the insurance company is $\$2000 \times 1 + (-\$100,000) \times 1/50 = 0$, so they can expect to lose money if they charge less than $2000 per policy.

6. **a**. The shortcut saves 5 minutes 90% of the time, but it loses 20 minutes 10% of the time.

7. **c**. The results of the previous five games have no bearing on the outcome of the next game.

8. **c**. The results of the previous five games have no bearing on the next game.

9. **c**. In order to compute the probability of winning on any single trial, you would need to know the probability of winning on each of the various payouts.

10. **c**. 97% of $10 million is $9.7 million, which leaves $300,000 in profit for the casino (over the long haul).

DOES IT MAKE SENSE?

7. Makes sense. An organization holding a raffle usually wants to raise money, which means the expected value for purchasers of raffle tickets will be negative.

9. Does not make sense. Each of the 16 possible outcomes when four coins are tossed is equally likely.

11. Does not make sense. The slot machine doesn't remember what happened on the previous 25 pulls, and your probability of winning remains the same for each pull.

BASIC SKILLS AND CONCEPTS

13. You should not expect to get exactly 5000 heads (the probability of that happening is quite small), but you should expect to get a result near to 5000 heads. The law of large numbers says that the proportion of a large number of coin tosses that result in heads is near to the probability of getting heads on a single toss. Since the probability of getting heads is 1/2, you can expect that around one-half of the 10,000 tosses will result in heads.

15. The probability of tossing three heads is 1/8, and that of not tossing three heads is 7/8. Thus the expected value of the game is $6 × 1/8 + (–$1) × 7/8 = –$0.125. Though one can assign a probability to the outcome of a single game, the result cannot be predicted, so you can't be sure whether you will win or lose money in one game. However, over the course of 100 games, the law of averages comes into play, and you can expect to lose about $12.50.

17. The probability of rolling two even numbers is 9/36 = 1/4 (of the 36 possible outcomes, 9 are "two even numbers"), and the probability of not rolling two even numbers is 27/36 = 3/4. Thus the expected value is $2 × 1/4 + (–$1) × 3/4 = –$0.25. The outcome of one game cannot be predicted, though over 100 games, you should expect to lose.

19. The expected value for a single policy is $1000 × 1 + (–$10,000) × 1/50 + (–$25,000) × 1/100 + (–$60,000) × 1/250 = $310. The company can expect a profit of $310 × 10,000 = $3.1 million.

21. For the sake of simplicity, assume that you arrive on the half-minute for each minute of the hour (that is, you arrive 30 seconds into the first minute of the hour, 30 seconds into the second minute of the hour, etc.). The probability that you arrive in any given minute in the first half-hour is 1/30, and your wait times (in minutes) for the first 30 minutes are 29.5, 28.5, 27.5, … , 0.5. Thus your expected wait time is 29.5 × 1/30 + 28.5 × 1/30 + 27.5 × 1/30 + … + 0.5 × 1/30 = 15 minutes. Of course the same is true for the second half-

hour. (Note for instructors: if you wanted to allow for arrival at *any* time during a given minute, you would need to use calculus in order to solve the problem with any level of rigor, because the wait time function is continuous.)

23. a. Heads came up 46% of the time, and you've lost $8.

b. Yes, the increase in heads is consistent with the law of large numbers: the more often the coin is tossed, the more likely the percentage of heads is closer to 50%. Since 47% of 300 is 141, there have been 141 heads and 159 tails, and you are now behind $18.

c. You would need 59 heads in the next 100 tosses (so that the total number of heads would be 200) in order to break even. While it's possible to get 59 heads in 100 tosses, a number nearer to 50 is more likely.

d. If you decide to keep playing, the most likely outcome is that half of the time you'll see heads, and half the time you'll see tails, which means you'll still be behind, overall .

25. a. If you toss a head, the difference becomes 15; if you toss a tail, the difference becomes 17.

b. The most likely outcome after 1000 more flips is 500 heads and 500 tails (in which case the difference would remain at 16). The distribution of the number of heads is symmetric so that any deviation from "500 heads/500 tails" is as likely to be in the upper tail as the lower tail.

c. Once you have fewer heads than tails, the most likely occurrence after additional coin tosses is that the deficit of heads remains.

27. The probability of getting ten heads in ten tosses is $(1/2)^{10} = 1/1024$. There are many ten-toss sequences in the 1000 tosses (991 to be precise), and you should expect, on average, one streak of ten heads in every 1024 ten-toss sequences.

29. a. Suppose you bet $1. Then your expected value is $1 × 0.493 + (–$1) × 0.507 = –$0.014. In general, the expected value is –1.4% of your bet, which means the house edge is 1.4% (or 1.4¢ per dollar bet).

b. You should expect to lose 100 × $0.014 = $1.40.

c. You should expect to lose $500 × 1.4% = $7.

d. The casino should expect to earn 1.4% of $1 million, which is $14,000.

FURTHER APPLICATIONS

31. The expected value is $(-\$1) \times 1 + \$30,000,000 \times 1/175,223,510 + \$1,000,000 \times 1/5,153,633 + \$10,000 \times 1/648,976 + \$100 \times 1/19,088 + \$100 \times 1/12,245 + \$7 \times 1/360 + \$7 \times 1/706 + \$4 \times 1/111 + \$4 \times 1/55 = -\$0.47$. You can expect to lose $365 \times \$0.47 = \$172$.

33. The expected value is $(-\$1) \times 1 + \$15,000,000 \times 1/175,711,536 + \$250,000 \times 1/3,904,701 + \$10,000 \times 1/689,065 + \$150 \times 1/15,313 + \$150 \times 1/13,781 + \$10 \times 1/844 + \$7 \times 1/306 + \$3 \times 1/141 + \$2 \times 1/75 = -\$0.73$. You can expect to lose $365 \times \$0.73 = \$266$.

35. a. $23,325/23,684 \approx 0.9848$.

   b. $344/718 \approx 0.4791$.

   c. The expected value for 1-point conversions is $1 \times 0.9848 + 0 \times 0.0152 = 0.9848$ points. The expected value for 2-point conversations is $2 \times 0.4791 + 0 \times 0.5209 = 0.9582$ points.

   d. In most cases, it makes sense to take the almost certain extra-point, rather than risk a failed 2-point attempt. One situation where most coaches will go for a 2-point attempt: if your team is down by 8 points late in the game, and you score a touchdown (6 points), you don't benefit much by taking the extra point. You'll still need a field goal (3 points) to win the game. However, if you go for a 2-point attempt, you may tie the game, and send it into overtime. If you miss the 2-point attempt, you'll still be able to win the game with a field goal.

37. The expected value for the number of people in a household is $1.5 \times 0.57 + 3.5 \times 0.32 + 6 \times 0.11 = 2.6$ people. The expected value is a weighted mean, and it is a good approximation to the actual mean as long as the midpoints are representative of the categories.

UNIT 7D

QUICK QUIZ

1. **a.** The accident rate will go down because the numerator of the "accident rate per mile" ratio is decreasing while the denominator is increasing.

2. **c.** The blue curve in the graph is rising as time progresses.

3. **b.** Though there has been some variation in the number of deaths, the number of miles driven has steadily increased while the number of deaths have decreased, and thus the death rate has generally decreased.

4. **c.** The death rate per person is the number of deaths due to AIDS divided by the population.

5. **c.** The death rate per person is $\dfrac{73,300}{315 \text{ million}}$, and this needs to be multiplied by 100,000 to get the death rate per 100,000 people.

6. **a.** The chart shows that children in their younger years have a significant risk of death.

7. **c.** Your life expectancy decreases as you age because the number of years before you die is decreasing.

8. **c.** Current life expectancies are calculated for current conditions – if these change, life expectancies can also change.

9. **b.** The life expectancies of men are consistently below those of women (and from the graph, the gap has widened a bit in the last several decades).

10. **a.** If life expectancy goes up, people will live longer, and will draw Social Security benefits for a longer period of time.

DOES IT MAKE SENSE?

5. Does not make sense. Automobiles are successfully sold in huge quantities despite the risks involved.

7. Does not make sense. Your life expectancy is an average life expectancy for people born when you were. Much more important is your genetic makeup, and the choices you make throughout life (your diet, your line of work, the activities you choose to participate in, etc.).

BASIC SKILLS AND CONCEPTS

9. The per-mile death rate was $\dfrac{32,885}{3.0 \times 10^{12}} = 1.05 \times 10^{-8}$. Multiply by 100 million to get 1.05 deaths per 100 million vehicle-miles.

11. The per-driver rate was $\dfrac{32,885}{2.1 \times 10^{8}} = 1.57 \times 10^{-4}$. Multiply by 100,000 to get 15.7 fatalities per 100,000 drivers.

13. In 2006 there were $\dfrac{1523}{2.4 \times 10^7} \times 10^5 = 6.3$ accidents per 100,000 hours flown. In 2010 there were $\dfrac{1384}{2.02 \times 10^7} \times 10^5 = 6.9$ accidents per 100,000 hours flown. The rates are fairly close, but it seems flying was a little less safe in 2010.

15. The probability of death by diabetes is $\dfrac{73,300}{3.15 \times 10^8} = 0.00023$. The probability of death by kidney disease is $\dfrac{45,700}{3.15 \times 10^8} = 0.00015$. Death by diabetes is about $\dfrac{0.00023}{0.00015} = 1.53$ times more likely than death by kidney disease.

17. The death rate due to Alzheimer's is about $\dfrac{84,700}{3.15 \times 10^8} \times 10^5 = 26.9$ deaths per 100,000 people.

19. The death rate due to stroke is about $\dfrac{128,900}{3.15 \times 10^8} \times 10^5 = 40.92$ deaths per 100,000 people, so in a city of 500,000, you would expect $40.92 \times 5 = 205$ people to die from a stroke each year.

21. The death rate for 60-year-olds is about 9 per 1000, and for 65-year-olds, it's about 13 per 1000, so the death rate for the category of 60- to 65-year-olds is about 11 deaths per 1000 people. Multiply 11/1000 by 13.4 million to find the number of people in this category expected to die in a year: $\dfrac{11}{1000} \times 13.4 \times 10^6 \approx 147,400$ people.

(Note: Answers will vary depending on values read from the graphs)

23. The life expectancy for a 40-year-old is about 40 years, so the average 40-year-old can expect to live to age 80.

25. The death rate for a 50-year-old is about 5 per 1000, which means the probability that a randomly selected 50-year-old will die in the next year is 5/1000. Those that don't die will file no claim; the beneficiaries of those that do will file a $50,000 claim. Thus the expected value of a single policy (from the point of view of the insurance company) is $200 × 1 + (–$50,000) × 5/1000 + ($0) × 995/1000 = –$50. The company insures 1 million 50-year-olds, and expects to lose $50 per policy, so it can expect a loss of $50 million.

27. Life expectancy for women increased by $\dfrac{80 - 48}{48} = 0.6667 = 66.67\%$, so the life expectancy for women in 2100 would be $80 \times 1.6667 = 133$ years. Answers will vary, but it appears the prediction in Example 5 seems more reasonable.

FURTHER APPLICATIONS

29. a. In Utah, about $\dfrac{52,258}{365} = 143$ people were born per day.

    b. In Maine, about $\dfrac{12,970}{365} = 36$ people were born per day.

    c. The birth rate in Utah was $\dfrac{52,258}{2.8 \times 10^6} \times 10^5 = 1866$ births per 100,000 people.

    d. The birth rate in Maine was $\dfrac{12,970}{1.3 \times 10^6} \times 10^5 = 998$ births per 100,000 people.

31. a. There were about $\dfrac{13}{1000} \times 309$ million = 4.02 million births in the U.S.

    b. There were $\dfrac{7.4}{1000} \times 309$ million = 2.29 million deaths in the U.S.

    c. The population increased by 4.02 million – 2.29 million = 1.73 million people.

    d. About 3.5 million – 1.73 million = 1.77 million people immigrated to the U.S. (This estimate neglects to account for emigration, though the number of people who emigrate from the U.S. is relatively low.) The proportion of the overall population growth due to immigration was 1.77/3.5 = 51%.

33. a. The probability of giving birth to more than one baby is about (133,000 + 5500 + 300) ÷ 4,000,000 = 0.035.

    b. We first need to calculate the total number of children born each year. There were 4,000,000 – 133,000 – 5500 – 300 = 3,861,200 single births, so there will be a total of 3,861,200 + 2 × 133,000 + 3 × 5500 + 4 × 300 = 4,144,900 children born each year. The probability that a randomly selected newborn is a twin is about (2 × 133,000) ÷ 4,144,900 = 0.064.

33. (continued)

   c. The probability that a randomly selected newborn is a twin, a triplet, or a quadruplet is about $(2 \times 133{,}000 + 3 \times 5500 + 4 \times 300) \div 4{,}144{,}900 = 0.068$.

UNIT 7E

QUICK QUIZ

1. **b.** This is an arrangement with repetition, with 36 choices at each selection.

2. **c.** Assuming each person gets one entrée, this is a permutation of 3 items taken 3 at a time, and $_3P_3 = 3!/0! = 6$.

3. **c.** The order of selection is important when counting the number of ways to arrange the roles.

4. **b.** Once the five children have been chosen, we are arranging just those five (and all five) into the various roles, and thus this is a permutation of 5 items taken 5 at a time, or $_5P_5$.

5. **b.** The number of permutation of n items taken r at a time is always at least as large as the number of combinations of n items taken r at a time (by a factor of $r!$), and $_{15}P_7 = 32{,}432{,}400$ is much larger than $_7P_7 = 5040$.

6. **c.** The variable n stands for the number of items from which you are selecting, and the variable r stands for the number of items you select. Here, $n = 9$, and $r = 4$, and thus $(n - r)! = 5!$.

7. **c.** The number of combinations of 9 items taken 4 at a time is $_9C_4 = \dfrac{9!}{(9-4)!4!} = \dfrac{9 \times 8 \times 7 \times 6}{4 \times 3 \times 2 \times 1}$.

8. **b.** If there is a drawing, the probability that one person is selected is 100%.

9. **c.** The probability that it will be you is $1/100{,}000 = 0.00001$, so the probability it will not be you is $1 - 0.00001 = 0.99999$.

10. **c.** As shown in Example 8b, the probability is about 57%.

DOES IT MAKE SENSE?

5. Makes sense. The permutations formula is used when order of selection is important.

7. Makes sense. The number of such batting orders is $_{25}P_9 = 7.4 \times 10^{11}$, which is nearly 1 trillion ways. Clearly there is no hope of trying all of them.

9. Makes sense. This illustrates the general principle that *some* coincidence is far more likely that a *particular* coincidence.

BASIC SKILLS AND CONCEPTS

11. $6! = 720$

13. $5! \div 3! = 20$

15. $\dfrac{12!}{4!3!} = 3{,}326{,}400$

17. $\dfrac{11!}{3!(11-3)!} = 165$

19. $\dfrac{8!}{3!(8-3)!} = 56$

21. $\dfrac{6!8!}{4!5!} = 10{,}800$

23. This is an arrangement with repetition, where we have 10 choices (the digits 0–9) for each of 9 selections, and thus there are $10^9 = 100{,}000{,}000$ possible zip codes.

25. This is a permutation of 26 items taken 5 at a time because the order in which the characters are arranged is important (a different order results in a different password). Thus there are $_{26}P_6 = 165{,}765{,}600$ passwords.

27. This is a permutation of 8 items taken 8 at a time, and $_8P_8 = 40{,}320$.

29. Because order of selection does not matter, this is a combination of 15 items taken 6 at a time, or $_{15}C_6 = 5005$.

31. This is an arrangement with repetition, where there are 2 choices for each of 5 selections, and thus there are $2^5 = 32$ birth orders.

33. This is an arrangement with repetition, so use the multiplication principle to find that there are $26^2 \times 10^5 = 67{,}600{,}000$ such license plates.

35. This is an arrangement with repetition, so use the multiplication principle to get $9 \times 9 \times 9 \times 9 \times 9 \times 9 \times 10 \times 10 \times 10 \times 10 = 5{,}314{,}410{,}000$.

37. If we assume that the letters can be repeated, this is an arrangement with repetition, and there are 4 choices for each of the 3 selections, which implies there are $4^3 = 64$ different "words."

39. The order in which the songs appear matters, so this is a permutation of 12 objects taken 12 at a time, and thus there are $_{12}P_{12} = 479{,}001{,}600$ different ways to order the songs.

41. Following Example 8 in the text, the probability that no one has your birthday is $1 - \left(\dfrac{364}{365}\right)^{11} \approx 0.0297$. The probability that at least one pair shares a birthday is given by

$$1 - \frac{364 \times 363 \times \ldots \times 354}{365^{11}} = 1 - \frac{1.276 \times 10^{28}}{1.532 \times 10^{28}} \approx 0.167 \, .$$

FURTHER APPLICATIONS

43. a. Use the multiplication principle to get $20 \times 8 = 160$ different sundaes.

 b. This is an arrangement with repetition, so there are $20^3 = 8000$ possible triple cones.

 c. Since you specify the locations of the flavors, the order matters, and you should use permutations. There are $_{20}P_3 = 6840$ possibilities.

 d. The order of the flavors doesn't matter, so use combinations. There are $_{20}C_3 = 1140$ possibilities.

45. This problem is best solved by trial and error. For Luigi, we know that 56 is the number of combinations of n items taken 3 at a time, and thus $_{n}C_3 = 56$. After a little experimentation, you will discover that $n = 8$; that is, Luigi uses 8 different toppings to create 56 different 3-topping pizzas. In a similar fashion, you can discover that Ramona uses 9 different toppings (solve $_{n}C_2 = 36$ for n).

47. There are $_{36}C_6 = 1{,}947{,}792$ ways to choose the balls, and only one way to match all six numbers, so the probability is $1/1{,}947{,}792$.

49. There are $_{52}C_5 = 2{,}598{,}960$ five-card hands. Four of those consist of the 10, J, Q, K, and A of the same suit, so the probability is $4/2{,}598{,}960 = 1/649{,}740$.

51. There are $_{20}C_3 = 1140$ ways to choose the top three spellers, and only one way to guess the correct three, so the probability is $1/1140$.

53. There are $_{52}C_5 = 2{,}598{,}960$ five-card hands. In order to get four-of-a-kind in, say, aces, you need to select all four aces, and then any of the other remaining 48 cards. Thus there are 48 ways to get a four-of-a-kind in aces. This is true for the other 12 card values as well, so there are $13 \times 48 = 624$ different four-of-a-kind hands. The desired probability is then $624/2{,}598{,}960 \approx 0.00024$.

55. a. The probability of any one individual winning five games in a row is $0.48^5 = 0.025$, assuming you play only five games. It is considerably higher if you play numerous games. If 2000 people each play 5 games in a row, we would expect about $2000 \times 0.025 = 50$ people to have such a "hot streak."

 b. The probability of any one individual winning ten games in a row is $0.48^{10} = 0.00065$, assuming you play only ten games. If 2000 people each play 10 games in a row, we would expect about $2000 \times 0.00065 = 1.3$ people to have such a "hot streak."

57. There are $_{56}C_5 = 3{,}819{,}816$ ways to choose the first five numbers and 46 ways to choose Megaball number, so there are $3{,}819{,}816 \times 45 = 175{,}711{,}536$ ways to play. There is only one way to win the jackpot, so the probability is $1/175{,}711{,}536$ or about 0.000000006.

TECHNOLOGY EXERCISES

63. a. There are $_{44}C_5 = 1{,}086{,}008$ ways to choose the balls, and only one way to match all five numbers, so the probability is $1/1{;}086{,}008$.

 b. There are $_{40}C_6 = 3{,}838{,}380$ ways to choose the balls, and only one way to match all six numbers, so the probability is $1/3{,}838{,}380$.

 c. The first lottery offers a better chance of winning.

UNIT 8A

QUICK QUIZ

1. **b.** The absolute change in the population was 10,000 people, and if population grows linearly, the absolute change remains constant. Thus the population at the end of the second year will be 110,000 + 10,000 = 120,000.

2. **c.** Because the population increases exponentially, it undergoes constant percent change. Its population increases by 10% in the first year, and thus at the end of the second year, the population will be 110,000 × 10% + 110,000 = 121,000.

3. **a.** Because your money is growing exponentially, if it doubles in the first 6 months, it will double every 6 months.

4. **b.** The absolute change in the number of songs is 200, and this occurred over the span of 3 months. Because the number of songs is increasing linearly, it will grow by 200 every 3 months, and it will take 6 months to grow to 800 songs.

5. **c.** Exponential decay occurs whenever a quantity decreases by a constant percentage (over the same time interval).

6. **c.** Based on the results of Parable 1 in the text, the total number of pennies needed is $2^{64} - 1 = 1.8 \times 10^{19}$, which is equivalent to $\$1.8 \times 10^{17}$. The federal debt is about \$17 trillion $\$\left(17 \times 10^{12}\right)$, which differs from the amount of money necessary by about a factor of 10,000.

7. **b.** Note that at 11:01, there are $2 = 2^1$ bacteria; at 11:02 there are $4 = 2^2$ bacteria; at 11:03 there are $8 = 2^3$, and so on. The exponent on the base of 2 matches the minute, and this pattern continues through 11:30, which means there are 2^{30} bacteria at that time.

8. **b.** At 11:31, there are 2^{31} bacteria, and as shown in Exercise 7, there were 2^{30} bacteria at 11:30. The difference, $2^{31} - 2^{30} = 2^{30}$, which is the number of bacteria added over that minute.

9. **a.** As with the population, the volume of the colony doubles every minute, so it only takes one minute for the volume to double from 1 to 2 m^3.

10. **b.** The growth of a population undergoing exponential change is comparatively slow in the initial stages, and then outrageously fast in the latter stages. This makes it difficult for members of the population to see that the growth they are currently experiencing is about to explode in short order.

DOES IT MAKE SENSE?

5. Makes sense. Any quantity that undergoes constant, positive percent change grows exponentially.

7. Makes sense. Any quantity that grows exponentially will eventually become very large, so as long as the small town's rate of growth is large enough, it can become a large city within a few decades.

BASIC SKILLS AND CONCEPTS

9. The population is growing linearly because its absolute change remains constant at 300 people per year. The population in four years will be 2500 + 4 × 300 = 3700.

11. This is exponential growth because the cost of food is increasing at a constant rate of 30% per month. Each month, the cost increases by a factor of 1.30, so your food bill would be R$100(1.30)^4 = R\$285.61$.

13. The price is decreasing exponentially because the percent change remains at a constant −14% per year. Each year, the price decreases by a factor of 0.86, so its price in three years will be $\$50(0.86)^3 = \$31.80$.

15. The house's value is increasing linearly because the price increases by the same amount each year. In five years, it will be worth \$100,000 + 5 × \$2,000 = \$110,000.

17. There should be $2^{15} = 32,768$ grains placed on the 16th square. The total number of grains would be $2^{16} - 1 = 65,535$, and it would weigh 65,535 × (1/7000) = 9.36 pounds.

19. When the chessboard is full, there are $2^{64} - 1$ grains, with total weight of $\left(2^{64} - 1\right) \times (1/7000) = 2.64 \times 10^{15}$ pounds, or about 1.3×10^{12} tons.

21. After 22 days, you would have $2^{22} = 4,194,304$ pennies, or $41,943.04.

23. After experimenting with various values of n in 2^n, you'll find that the balance has grown to about $1.4 billion after 37 days. 25. At 11:50, there are $2^{50} = 1.1 \times 10^{15}$ bacteria in the bottle. The bottle is full at 12:00 noon, so it was half-full at 11:59, 1/4-full at 11:58, 1/8-full at 11:57, and so on. Continuing in this manner, we find the bottle was 1/1024-full at 11:50. This could also be expressed as $1/2^{10}$-full at 11:50.

27. As discussed in the text, the volume of the colony of 2^{120} bacteria would occupy a volume of $1.3 \times 10^{15} \text{ m}^3$. Divide this by the surface area of the earth to get an approximate depth:

$$1.3 \times 10^{15} \text{ m}^3 \div 5.1 \times 10^{14} \text{ m}^2 = 2.5 \text{ m}.$$

This is quite a bit more than knee-deep (it's more than 8 feet).

FURTHER APPLICATIONS

29. a. (Only 100-year intervals are shown here).

Year	Population
2000	6.0×10^9
2100	2.4×10^{10}
2200	9.6×10^{10}
2300	3.8×10^{11}
2400	1.5×10^{12}
2500	6.1×10^{12}
2600	2.5×10^{13}
2700	9.8×10^{13}
2800	3.9×10^{14}
2900	1.6×10^{15}
3000	6.3×10^{15}

b. In the year 2800, there would be 3.9×10^{14} people, each occupying slightly more than one square meter, and in 2850, there would be 7.9×10^{14} people (each occupying slightly less than one square meter). Since $5.1 \times 10^{14} \text{ m}^2$ is in between these two values, it would happen between 2800 and 2850.

c. We would reach the limit when the earth's population reached $5.1 \times 10^{14} \div 10^4 = 5.1 \times 10^{10}$ people. This would happen shortly after the year 2150.

d. If by 2150 we've already reached the limit of 10^4 m^2 per person on earth, it will take only three more doubling periods to increase our need for space eight-fold. Since the doubling period is 50 years, this would take only 150 years, at which point we would have already exceeded the additional space available in the solar system (there is only five times more area out there).

31. a. and b.

Month	March 2010	March 2011	March 2012
Active users	431	680	901
Absolute change	$431 - 197$ $= 234$	$680 - 431$ $= 249$	$901 - 680$ $= 221$
Percent change	$\dfrac{431-197}{197}$ $= 118.8\%$	$\dfrac{680-431}{431}$ $= 57.8\%$	$\dfrac{901-680}{680}$ $= 32.5\%$

c. The absolute change is roughly constant and the percent change is decreasing, so the growth is closer to linear.

UNIT 8B

QUICK QUIZ

1. **a.** The factor by which an exponentially growing quantity grows is $2^{t/T_{\text{double}}}$.

2. **b.** The approximate doubling time formula says that a quantity's doubling time is about $70/r$, where r is the growth rate.

3. **a.** The approximate doubling time formula does not do a good job when the growth rate r goes higher than 15%.

4. **b.** The approximate doubling time formula, $T_{\text{double}} = 70/r$, can be rearranged to say $r = 70/T_{\text{double}}$.

5. **c.** During the first 12 years, half of the tritium decays, and during the second 12 years, half of that, or 1/4 decays.

6. **b.** Note that 2.8 billion years/4 = 700 million years. This implies the uranium-235 decayed by a factor of 2 four times, which means $1/2^4 = 1/16$ of the uranium-235 remains.

7. **b.** The half-life can be approximated by $T_{half} = 70/r$.

8. **c.** A property of logarithms states $\log_{10}(10^x) = x$.

9. **c.** The decimal equivalent of 20% is 0.2, though you should set $r = -0.2$ because the population is decreasing.

10. **a.** The doubling time formula states $T_{double} = \dfrac{\log_{10} 2}{\log_{10}(1+r)}$.

DOES IT MAKE SENSE?

9. Does not make sense. In 50 years, the population will increase by a factor of 4.

11. Does not make sense. While it's true that half will be gone in ten years, over the next ten years, only half of what remains will decay, which means there will still be 1/4 of the original amount.

BASIC SKILLS AND CONCEPTS

13. True. $10^0 = 1$, and $10^1 = 10$, so $10^{0.928}$ should be between 1 and 10 (because 0.928 is between 0 and 1).

15. False. $10^{-5.2} = \dfrac{1}{10^{5.2}}$, which is positive.

17. False. While it is true that π is between 3 and 4, $\log_{10} \pi$ is not (it's between 0 and 1).

19. False. $\log_{10}(10^6) = 6$, and $\log_{10}(10^7) = 7$. Because 1,600,000 is between 10^6 and 10^7, $\log_{10} 1,600,000$ is between 6 and 7.

21. True. Since $\log_{10}\left(\dfrac{1}{4}\right) = \log_{10}\left(4^{-1}\right) = -\log_{10} 4$, we know $\log_{10}\left(\dfrac{1}{4}\right)$ is between –1 and 0 because $\log_{10} 4$ is between 0 and 1.

23. a. $\log_{10} 8 = \log_{10}(2^3) = 3\log_{10} 2 = 3 \times 0.301 = 0.903$

   b. $\log_{10} 2000 = \log_{10} 2 + \log_{10} 10^3 = 0.301 + 3 = 3.301$

   c. $\log_{10} 0.5 = \log_{10} 1 - \log_{10} 2 = 0 - 0.301 = -0.301$

   d. $\log_{10} 64 = \log_{10} 2^6 = 6\log_{10} 2 = 6 \times 0.301 = 1.806$

   e. $\log_{10}\left(\dfrac{1}{8}\right) = \log_{10} 2^{-3} = -3\log_{10} 2$
   $= -3 \times 0.301 = -0.903$

   f. $\log_{10} 0.2 = \log_{10}\left(\dfrac{2}{10}\right) = \log_{10} 2 - \log_{10} 10$
   $= 0.301 - 1 = -0.699$

25. In 24 hours, the fly population will increase by a factor of $2^{24/8} = 2^3 = 8$. Since one week is 168 hours, the population will increase by a factor of $2^{168/8} = 2^{21} = 2,097,152$ in one week.

27. The population will increase by a factor of four (quadruple) in 32 years because $2^{32/16} = 2^2 = 4$.

29. In 12 years, the population will be $12,000 \times 2^{12/10} \approx 28,000$. In 24 years, it will be $12,000 \times 2^{24/10} \approx 63,000$.

31. Note that 2 years is 24 months, and 4 years is 48 months. In 2 years, the number of cells will be $1 \times 2^{24/2.5} = 776$. In 4 years, the number will be $1 \times 2^{48/2.5} = 602,249$.

33. The year 2023 is 10 years later than 2013, so the population will be $7.1 \times 2^{10/45} = 8.3$ billion. In 2063 (50 years later), the population will be $7.1 \times 2^{50/45} = 15.3$ billion. In 2113 (100 years later), it will be $7.1 \times 2^{100/45} = 33.1$ billion.

35.

Month	Population
0	100
1	$100 \times (1.07)^1 = 107$
2	$100 \times (1.07)^2 = 114$
3	$100 \times (1.07)^3 = 123$
4	$100 \times (1.07)^4 = 131$
5	$100 \times (1.07)^5 = 140$
6	$100 \times (1.07)^6 = 150$
7	$100 \times (1.07)^7 = 161$
8	$100 \times (1.07)^8 = 172$
9	$100 \times (1.07)^9 = 184$
10	$100 \times (1.07)^{10} = 197$
11	$100 \times (1.07)^{11} = 210$
12	$100 \times (1.07)^{12} = 225$
13	$100 \times (1.07)^{13} = 241$
14	$100 \times (1.07)^{14} = 258$
15	$100 \times (1.07)^{15} = 276$

Because the population after ten months is 197, which is almost twice the initial population of 100, the doubling time is just over 10 months. The approximate doubling time formula claims that T_{double} = 70/7 = 10 months, which is pretty close to what we can discern from the table. Neither answer is exact, though.

37. $T_{double} \approx 70/3 = 23.3$ years. Prices will increase by a factor of $(1.03)^3 = 1.09$ in three years. Using the doubling time formula, prices would increase by a factor of $2^{3/23.3} = 1.09$, so the approximate doubling time formula does a reasonable job.

39. $T_{double} \approx 70/0.8 = 87.5$ months. Prices will increase by a factor of $(1.008)^{12} = 1.10$ in one year, and by a factor of $(1.008)^{96} = 2.15$ in eight years. Using the doubling time formula, prices would increase by a factor of $2^{12/87.5} = 1.10$ in one year, and by a factor of $2^{96/87.5} = 2.14$ in eight years, so the approximate doubling time formula does a reasonable job.

41. After 100 years, the fraction of radioactive substance that remains is $\left(\frac{1}{2}\right)^{100/50} = \frac{1}{4} = 0.25$. After 300 years, the fraction remaining is $\left(\frac{1}{2}\right)^{300/50} = \frac{1}{64} = 0.015625$.

43. The concentration of the drug decreases by a factor of $\left(\frac{1}{2}\right)^{24/18} \approx 0.40$ after 24 hours, and by a factor of $\left(\frac{1}{2}\right)^{72/18} = \frac{1}{16} = 0.0625$ after 72 hours.

45. In 30 years, the population will be $1,000,000 \times \left(\frac{1}{2}\right)^{30/20} \approx 354,000$, and it will be $1,000,000 \times \left(\frac{1}{2}\right)^{70/20} \approx 88,000$ in 70 years.

47. In 150 days, $1 \times \left(\frac{1}{2}\right)^{150/77} = 0.26$ kg of cobalt will remain, and $1 \times \left(\frac{1}{2}\right)^{300/77} = 0.067$ kg of cobalt will remain after 300 days.

49. $T_{half} \approx 70/7 = 10$ years. The fraction of the forest that will remain after 50 years is $(1 - 0.07)^{50} = 0.027$. Using the half-life formula, the fraction that will remain is $\left(\frac{1}{2}\right)^{50/10} = \frac{1}{32} = 0.03$, so the approximate half-life formula is doing a reasonable job.

51. $T_{half} \approx 70/8 = 8.75$ years. After 50 years, the population will be $10,000(1 - 0.08)^{50} = 155$ elephants. Using the half-life formula, the population will be $10,000 \times \left(\frac{1}{2}\right)^{50/8.75} = 190$ elephants, so the approximate half-life formula does not do a reasonable job.

53. The approximate doubling time is $T_{double} \approx 70/12 = 5.8$ years. The exact doubling time is $\frac{\log_{10} 2}{\log_{10}(1 + 0.12)} \approx 6.12$ years. The price of a $500 item in four years will be $\$500 \times 2^{4/6.12} = \$786.54$. (Note that this answer is using the rounded value for the "exact" doubling time – if you use all the digits your calculator stores, you should get a price of $786.76).

55. The approximate doubling time is $T_{double} \approx 70/4 = 17.5$ years. The exact doubling time is
$$\frac{\log_{10} 2}{\log_{10}(1+0.04)} \approx 17.67$$ years. The nation's population in 30 years will be $100,000,000 \times 2^{30/17.67} = 324,404,283$. (Note that this answer is using the rounded value for the "exact" doubling time – if you use all the digits your calculator stores, you should get a population of 324,339,751).

FURTHER APPLICATIONS

57. The amount of plutonium found today would be
$$10^{13} \times \left(\frac{1}{2}\right)^{4.6\times10^9 / 24,000} \approx 0$$ tons. The reason there is no plutonium left is that 4.6 billion ÷ 24,000 is nearly 200,000, which means whatever amount of plutonium was originally present has undergone about 200,000 halving periods, and that's enough to reduce any amount imaginable to 0 (even if the entire mass of the earth were radioactive plutonium at the outset, it would have all decayed to other elements in 4.6 billion years).

59. $T_{double} \approx 70/6 = 11.67$ years. Using the doubling time formula, emissions would increase by a factor of $2^{40/11.67} = 10.76$. The exact formula for exponential growth estimates that emissions will increase by a factor of $(1.06)^{40} = 10.29$ from 2010 to 2050.

UNIT 8C

QUICK QUIZ

1. **c.** Convert 75 million people per year into people per minute:
$$\frac{75\times10^6 \text{ people}}{\text{yr}} \times \frac{1 \text{ yr}}{365 \text{ d}} \times \frac{1 \text{ d}}{24 \text{ h}} \times \frac{1 \text{ hr}}{60 \text{ min}} = 143$$
people per minute (or about 140 people/min).

2. **b.** The text states the population is 7.1 billion people in 2013, growing at 1.1% per year. This implies a 2050 population of $7.1\times10^9 (1.011)^{37} = 10.6$ billion.

3. **b.** The birth rate peaked in 1960, and has seen a significant decline since then, but it is not the primary reason for the rapid growth – the decreasing death rate has played the most important role over the last three centuries.

61. The approximate half life is $T_{half} \approx 70/3 = 23.33$ years. The exact half life is
$$-\frac{\log_{10} 2}{\log_{10}(1-0.03)} \approx 22.76$$ years. So we would expect the number of homicides to be reduced to 400 in about 23 years.

63. a. $1000 \text{ mb} \times \left(\frac{1}{2}\right)^{1/7} = 906$ mb

b. $1000 \text{ mb} \times \left(\frac{1}{2}\right)^{8.848/7} = 416$ mb

c. $\left(\frac{1}{2}\right)^{1/7} = 0.91$, so atmospheric pressure decreases by about 10% per kilometer.

TECHNOLOGY EXERCISES

69. a. $\log_{10} 50 \approx 1.698970$

b. $\log_{10} 5000 \approx 3.689970$

c. $\log_{10} 0.05 \approx -1.301030$

d. $\log_{10} 25 \approx 1.397940$

e. $\log_{10} 0.20 \approx -0.698970$

f. $\log_{10} 0.04 \approx -1.397940$

4. **b.** This answer gives the correct definition of carrying capacity.

5. **c.** Answers **a** and **b** would tend to decrease the carrying capacity, while developing a cheap source of energy would allow us to solve some of the barriers to population growth, such as clean water for all.

6. **a.** The bacteria in a bottle exhaust their resources, at which point the population collapses.

7. **c.** The logistic model looks like exponential growth at the outset, but then the growth rate declines, and the population levels out to a level at (or below) the carrying capacity.

8. **a.** The birth rate in 1950 was quite high, and with no change in the death rate, this would lead to rapid exponential growth.

9. **a.** The approximate doubling time of a population growing at 1.1% per year is 70/1.1 = 63.6 years.

10. **c.** Answers **a** and **b** would be required in order for the population to level out (the birth rate must decline to equal the death rate if population is to remain steady, and a population of 10 billion is 50% larger than the current level of 7.1 billion, and thus we would need an increase of 50% in food production). This leaves answer **c**, which is not necessary to sustain a population of 10 billion.

DOES IT MAKE SENSE?

7. Makes sense. If the current growth rate of 1.2% remains steady for the next decade, we can expect a population of $6.8 \times 10^9 (1.012)^{10} = 7.7$ billion, which is an increase of 900 million people (i.e. more than twice the size of the current U.S. population of 304 million).

9. Does not make sense. The carrying capacity depends upon many factors, such as our ability to produce food for the current population.

11. Does not make sense. Predator/prey models often include the idea of overshoot and collapse.

BASIC SKILLS AND CONCEPTS

13. $T_{\text{double}} \approx 70/0.9 = 78$ years. Under this assumption, the population in 2050 would be $7.1 \times 2^{37/78} \approx$ 9.9 billion.

15. $T_{\text{double}} \approx 70/1.6 = 44$ years. Under this assumption, the population in 2050 would be $7.1 \times 2^{37/44} \approx$ 12.7 billion.

17. **a.** The growth rate in 1980 was 51.8 – 24.1 = 27.7 per 1000 people. In 1995, it was 52.6 – 20.1 = 32.5 per 1000 people. In 2010, it was 42.3 – 15.1 = 27.2 per 1000 people.

    b. The birth rate and death rate both declined overall, which resulted in a constant positive growth rate. This implies a growing population. If these trends continue, we can expect an increasing population in Afghanistan, though because there are so many factors that influence the growth rate of a country, it would be unwise to put much faith in these predictions 20 years hence.

19. **a.** The growth rate in 1980 was 16.0 – 11.3 = 4.7 per 1000 people. In 1995, it was 10.9 – 14.6 = –3.7 per 1000 people. In 2010, it was 11.8 – 14.0 = –2.2 per 1000 people.

    b. The growth rate fell to a negative value, which results in a decreasing population. If these trends continue, we can expect a slightly declining or stable population in Russia, though because there are so many factors that influence the growth rate of a country, it would be unwise to put much faith in these predictions 20 years hence.

21. When the population is 10 million, the actual growth rate is $4\% \times \left(1 - \dfrac{10{,}000{,}000}{60{,}000{,}000}\right) \approx 3.3\%$.

    When the population is 30 million, the actual growth rate is $4\% \times \left(1 - \dfrac{30{,}000{,}000}{60{,}000{,}000}\right) \approx 2.0\%$.

    When the population is 50 million, the actual growth rate is $4\% \times \left(1 - \dfrac{50{,}000{,}000}{60{,}000{,}000}\right) \approx 0.67\%$.

FURTHER APPLICATIONS

23. $T_{\text{double}} \approx 70/0.7 = 100$ years. In 2050, the population will be $3.15 \times 10^8 \times 2^{37/100} = 407$ million. In 2100, the population will be $3.15 \times 10^8 \times 2^{87/100} = 576$ million.

25. $T_{\text{double}} \approx 70/1.0 = 70$ years. In 2050, the population will be $3.15 \times 10^8 \times 2^{37/70} = 454$ million. In 2100, the population will be $3.15 \times 10^8 \times 2^{87/70} = 746$ million.

27. Answers will vary with student's age. $T_{\text{double}} \approx 70/1.1 = 63.6$ years. For a student who was 17 in 2013: population at age 50 will be $7.1 \times 2^{(50-17)/63.6} = 10.2$ billion; at age 80 the population will be $7.1 \times 2^{(80-17)/63.6} = 14.1$ billion; at age 100 the population will be $7.1 \times 2^{(100-17)/63.6} = 17.5$ billion

29. We must first compute the base growth rate r in the logistic model, using the 1960s data:

$$r = \frac{2.1\%}{\left(1 - \dfrac{3 \text{ billion}}{8 \text{ billion}}\right)} = 3.36\%.$$

The current growth rate can now be predicted using a population of 7.1 billion:

$$\text{growth rate} = 3.36\% \times \left(1 - \frac{7.1 \text{ billion}}{8 \text{ billion}}\right) = 0.38\%.$$

This is lower than the actual current growth rate of 1.1%, which indicates the assumed carrying capacity of 8 billion is too low to fit well with the data from the 1960s to present.

31. We must first compute the base growth rate r in the logistic model, using the 1960s data:

$$r = \frac{2.1\%}{\left(1 - \dfrac{3\text{ billion}}{15\text{ billion}}\right)} = 2.63\%.$$

The current growth rate can now be predicted using a population of 7.1 billion:

growth rate $= 2.625\% \times \left(1 - \dfrac{7.1\text{ billion}}{15\text{ billion}}\right) = 1.38\%$.

This is higher than the actual current growth rate of 1.1%, which indicates the assumed carrying capacity of 15 billion is too high to fit well with the data from the 1960s to present.

UNIT 8D

QUICK QUIZ

1. **a.** Each unit on the magnitude scale represents about 32 times as much energy as the previous magnitude.

2. **b.** Most deaths in earthquakes are caused by the collapse of buildings, landslides, and tsunamis.

3. **a.** By definition, a sound of 0 dB is the softest sound audible by a human.

4. **b.** From the definition of the decibel scale, we have

$$\frac{\text{intensity of sound}}{\text{intensity of softest audible sound}} = 10^{(95/10)} = 10^{9.5}.$$

The ratio on the left tells us this sound has intensity $10^{9.5}$ times as large as the softest audible sound.

5. **a.** From the definition of the decibel scale, we have

$$\frac{\text{intensity of sound}}{\text{intensity of softest audible sound}} = 10^{(10/10)} = 10.$$

The ratio on the left tells us a 10-dB sound is ten times as large as the softest audible sound, which is 0 dB.

6. **c.** Gravity follows the inverse square law, so its strength is proportional to the square of the distance between the objects. Thus if this distance is tripled (multiplied by 3), the strength of the force of gravity is decreased by a factor of $3^2 = 9$.

7. **c.** On the pH scale, a lower number corresponds to higher acidity.

8. **b.** Because $[H^+] = 10^{-pH}$, we have $[H^+] = 10^{-0} = 1$, so the hydrogen ion concentration is 1 mole per liter.

33. With an annual growth rate of 2%, the approximate doubling time would be $T_{double} \approx 70/2 = 35$ years. The population of the city in the year 2020 (ten years beyond 2010) would have been about $100,000(2)^{10/35} = 122,000$. The predicted populations in 2030 and 2070 would be 149,000 and 328,000, respectively (the computation is similar to the 2020 prediction). With an annual growth rate of 5%, the approximate doubling time would be $T_{double} \approx 70/5 = 14$ years. The population of the city in the year 2020 would have been $100,000(2)^{10/14} = 164,000$. The predicted populations in 2030 and 2070 would be 269,000 and 1,950,000, respectively.

9. **c.** Because $pH = -\log_{10}[H^+]$, we have $pH = -\log_{10} 10^{-5} = -(-5) = 5$.

10. **a.** A lake damaged by acid rain has a relatively low pH. To bring it back to health, you want to raise the pH.

DOES IT MAKE SENSE?

5. Does not make sense. Each unit on the magnitude scale represents an increase of nearly 32 fold in the amount of energy released, and thus an earthquake of magnitude 8 releases about 32^4 (close to 1 million) times as much energy as an earthquake of magnitude 4. This does not imply that a magnitude 8 earthquake will do 1 million times as much damage, but it's not likely it will do only twice the damage of a magnitude 4 quake.

7. Does not make sense. The pH of water is related to its hydrogen ion concentration, which is not affected by its volume.

BASIC SKILLS AND CONCEPTS

9. $E = (2.5 \times 10^4) \times 10^{1.5 \times 6} = 2.5 \times 10^{13}$ joules

11. $E = (2.5 \times 10^4) \times 10^{1.5 \times 9} = 7.9 \times 10^{17}$ joules

13. The energy released by a magnitude 6 quake is 2.5×10^{13} joules (see Exercise 10), and since $\left(5 \times 10^{15}\right) \div \left(2.5 \times 10^{13}\right) = 200$, the bomb releases 200 times as much energy.

15. As shown in Table 8.5, ordinary conversation is around 60 dB, and is $10^6 = 1,000,000$ times as loud as the softest audible sound.

17. The loudness is $10\log_{10}(10\times10^6) = 70$ dB.

19. As with Example 4 in the text, we can compare the intensities of two sounds using the following:

$$\frac{\text{intensity of sound 1}}{\text{intensity of sound 2}} = 10^{(55-10)/10} = 31,623.$$

Thus a 55-dB sound is 31,623 times as loud as a 10-dB sound.

21. The distance has decreased by a factor of 3, which means the intensity of sound will increase by a factor of $3^2 = 9$.

23. The distance has decreased by a factor of 5, which means the intensity of sound will increase by a factor of $5^2 = 25$.

25. Each unit of increase on the pH scale corresponds to a decrease by a factor of 10 in the hydrogen ion concentration. Thus an increase of 4 on the pH scale corresponds to a decrease in the hydrogen ion concentration by a factor of $10^4 = 10,000$. This makes the solution more basic.

27. $[\text{H}^+] = 10^{-8.5} = 3.16\times10^{-9}$ mole per liter

29. $\text{pH} = -\log_{10}(0.001) = 3$. This is an acid, as is any solution with pH less than 7.

31. Acid rain (pH = 2) is $10^4 = 10,000$ times more acidic than ordinary water (pH = 6). See Exercise 25.

FURTHER APPLICATIONS

33. According to Table 8.4, earthquakes with magnitudes between 2 and 3 are labeled "very minor," and they occur about 1000 times per year worldwide. The glasses in the cupboards of the Los Angeles residents near the epicenter may rattle around a bit, but nothing serious would be expected to happen.

35. A solution with a pH of 12 is very basic: 100,000 times more basic than pure water. When children ingest such liquids, the mouth, throat, and digestive tracts are often burned. One would hope that a secondary effect of the incident would be that an adult would call the poison center immediately.

37. You probably would not be able to hear your friend (the noise of the sirens is louder). If your friend is shouting: "Watch out for that falling piano above you!" you might die.

39. a. The distance has increased by a factor of 100, so the intensity of the sound will decrease by a factor of $100^2 = 10,000$ times (this is due to the inverse square law). According to Table 8.5, busy street traffic is 80 dB, and is 10^8 times as loud as the softest audible sound. Note that a 40-dB sound is 10^4 times as loud as the softest sound, and thus is 10,000 times less intense than an 80-dB sound.

b. As with Example 4 in the text, we can compare the intensities of two sounds using the following:

$$\frac{\text{intensity of sound 1}}{\text{intensity of sound 2}} = 10^{(135-120)/10} = 31.6.$$

Thus a 135-dB sound is 31.6 times as loud as a 120-dB sound. In order to reduce the intensity of the sound by a factor of 31.6, you need to increase the distance from the speaker by a factor of $\sqrt{31.6} = 5.6$, so you should sit 5.6×10 m $= 56$ meters away.

c. The distance from you to the booth (8 meters) is 8 times as much as the distance between the people talking (1 meter), so the intensity of the sound will have decreased by a factor of 64 by the time it reaches you. A 20-dB sound is 100 times louder than the softest audible sound, and when decreased by a factor of 64, it becomes (100/64) times as loud as the softest sound. Therefore its loudness in decibels is $10\log_{10}\left(\dfrac{100}{64}\right) = 1.94$ dB.

Note that a 60-dB sound is 10,000 times as loud as a 20-dB sound. If you wanted to amplify the sound that reaches you to 60 dB, you would need to increase it by a factor of 64 to overcome the loss of intensity due to distance (this would bring it back to 20 dB), and increase it again by a factor of 10,000 to raise it to the 60-dB level. This is equivalent to amplifying the sound by a factor of 640,000. Pray that the waiter doesn't come to your table and say a loud "hello" while your earpiece is in.

41. a. $[H^+] = 10^{-4}$ mole per liter.

b. The lake water with pH = 7 has a hydrogen ion concentration of 10^{-7} mole per liter. There are 3.785 liters in a gallon, so a 100 million gallon lake has 378,500,000 = 3.785×10^8 liters in it. Multiplying the hydrogen ion concentration by the volume of the lake gives us 37.85 moles of hydrogen ions before the acid was added. In a similar fashion, we find that 100,000 gallons (378,500 liters) of acid with pH = 2 ($[H^+] = 10^{-2}$) has 3785 moles of hydrogen ions. Combining both the volumes of the liquids and the moles of hydrogen ions gives us a hydrogen ion concentration of $\dfrac{37.85 + 3785 \text{ mole}}{3.785 \times 10^8 + 378,500 \text{ L}} = 10^{-5}$ mole per liter. This implies the polluted lake has pH = $-\log_{10}(10^{-5}) = 5$.

c. Proceeding as in part **b**, the acid-rain lake (without added acid) has 37,850 moles of hydrogen ions. When the acid is added (increasing the number of moles by 3785), the hydrogen ion concentration becomes $\dfrac{37,850 + 3785 \text{ mole}}{3.785 \times 10^8 + 378,500 \text{ L}} =$ 1.1×10^{-4} mole per liter. This implies the polluted lake has pH = $-\log_{10}(1.1 \times 10^{-4}) = 3.96$.

d. The pollution could be detected if the acid was dumped into a lake with pH = 7 (part **b**), but it could not be detected if it was dumped into a lake with pH = 4 (part **c**).

UNIT 9A

QUICK QUIZ

1. **a.** A function describes how a dependent variable changes with respect to one or more independent variables.

2. **a.** The variable s is the independent variable, upon which the variable r depends.

3. **c.** The DJIA changes with respect to the passage of time, and thus is a function of time.

4. **b.** It is customary to plot the values of the independent variable along the horizontal axis.

5. **b.** The dependent variable is normally plotted along the vertical axis, and in this case, the dependent variable is z (with an independent variable of w).

6. **b.** The range of a function consists of the values of the dependent variable.

7. **a.** The speed of the car is the independent variable (because gas mileage changes with respect to speed), and the domain of a function is the set of values of interest for the independent variable.

8. **c.** The range of a function is the set of values corresponding to the values in the domain, which in turn are the set of values that make sense in the function. If you used a function to predict values outside the range, you would be forced to use values outside the domain; these values don't make sense.

9. **c.** Traffic volume on an urban freeway peaks in the morning rush hour, peaks again in the evening rush hour, and typically repeats this pattern day after day. If weekend traffic were significantly different from the weekday pattern, we would still see a weekly pattern repeating itself – this is a hallmark of a periodic function.

10. **b.** A model that does not agree well with past data probably won't do a good job of predicting the future.

DOES IT MAKE SENSE?

5. Makes sense. Climatologists use mathematical models to predict and understand the nature of the earth's climate.

7. Does not make sense. If your heart rate depends upon your running speed, heart rate is the dependent variable, and the range is used to describe the values of the dependent variable.

BASIC SKILLS AND CONCEPTS

9.

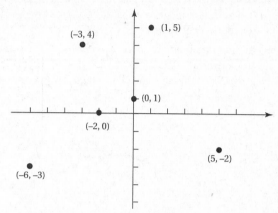

11. There is a functional relationship with *time* as the independent variable and *distance fallen* as the dependent variable.

13. There is a functional relationship with *price* as the independent variable and *demand* as the dependent variable.

15. As the volume of the tank increases, the cost of filling it also increases.

17. The temperature of the ocean decreases as latitude increases.

19. As the average speed of the car increases, the travel time decreases.

21. As the gas mileage increases, the cost of driving a fixed distance decreases (assuming that gas costs remain constant).

23. a. The pressure at 6000 feet is about 24 inches of mercury; the pressure at 18,000 feet is about 17 inches; and the pressure at 29,000 feet is about 11 inches.

    b. The altitude is about 8000 feet when the pressure is 23 inches of mercury; the altitude is about 14,000 feet when the pressure is 19 inches; and the altitude is about 25,000 feet when the pressure is 13 inches.

    c. It appears that the pressure reaches 5 inches of mercury at an altitude of about 50,000. The pressure approaches zero as one moves out of earth's atmosphere and into outer space.

25. a. The independent variable is time (in years), and the dependent variable is world population. The domain is the set of years from 1950 to 2000, and the range is the set of population values between roughly 2.5 billion and 6 billion people.

   b. The function shows a steadily increasing world population between 1950 and 2000.

27. a. The variables are (*time, temperature*), or (*date, temperature*). The domain is all days over the course of a year, and the range is the set of temperatures between 38°F and 85°F.

   b.

   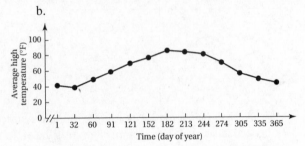

   c. The temperature increases during the first half of the year, and then decreases during the second half of the year.

29. a. The variables are (*speed, stopping distance*). The domain is speeds between 0 and 70 mi/hr, and the range is distances between 0 and 312 feet.

   b.

   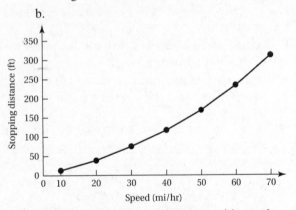

   c. The stopping distance increases with speed.

FURTHER APPLICATIONS

31. a. The domain of the function (*altitude, temperature*) is the set of altitudes of interest – say, 0 ft to 15,000 ft (or 0 m to 4000 m). The range is the set of temperatures associated with the altitudes in the domain; the interval 30°F to 90°F (or about 0°C to 30°C) would cover all temperatures of interest (this assumes a summer ascent of a mountain in the temperate zone of the earth).

b.

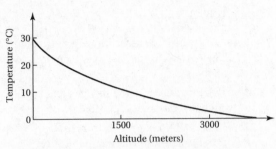

c. With some reliable data, this graph is a good model of how the temperature varies with altitude.

33. a. The domain of the function (*blood alcohol content, reflex time*) consists of all reasonable BACs (in gm/100 mL). For example, numbers between 0 and 0.25 would be appropriate. The range would consist of the reflex times associated with those BACs.

   b.

   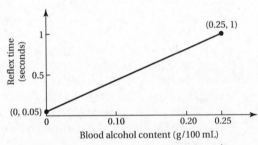

   c. The validity of this graph as a model of alcohol impairment will depend on how accurately reflex times can be measured.

35. a. The domain of the function (*time of day, traffic flow*) consists of all times over a full day. The range consists of all traffic flows (in units of cars per minute) at the various times of the day. We would expect light traffic flow at night, medium traffic flow during the midday hours, and heavy traffic flow during the two rush hours.

   b.

   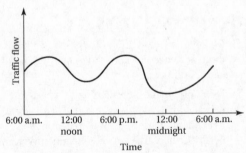

35. (continued)

    c. We would expect light traffic flow at night, medium traffic flow during the midday hours, and heavy traffic flow during the two rush hours. The graph of this function would be a good model only if based on reliable data.

37. a. The domain of the function (*number of people, number of handshakes*) consists of natural numbers from 2 (the minimum for a handshake to occur) to any arbitrary upper limit (for example, 10 people). The range consists of the number of handshakes, from 1 for two people up to the number for the upper limit of the domain. (Note: The formula for the number of handshakes between *n* people is $\dfrac{n(n-1)}{2}$.

    b.

    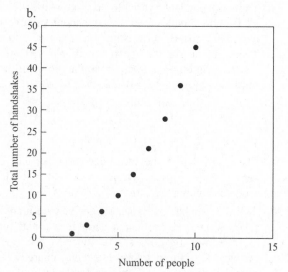

    c. The model is exact.

UNIT 9B

QUICK QUIZ

1. **c.** Constant absolute change over the same time interval (or other variable) gives rise to linear models, and this amounts to constant slope.

2. **b.** The slope of a linear function is the rate of change in the dependent variable with respect to the independent variable.

3. **c.** When a line slopes downward, the ratio of its change in *y* to change in *x* is negative, and this amounts to a negative slope or negative rate of change.

39. a. The domain of the function (*time, population of China*) is all years from 1900 to, say, 2010. The range consists of the population values of China during those years (roughly 400 million to 1.3 billion).

    b.

    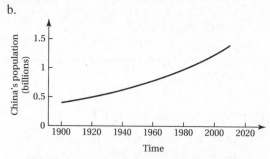

    c. With accurate yearly data, this graph would be a good model of population growth in China.

41. a. The domain of the function (*angle of cannon, horizontal distance traveled by cannonball*) is all cannon angles between $0°$ and $90°$. The range would consist of all horizontal distances traveled by the cannonball for the various angles in the domain. It is well known that a projectile has maximum range when the angle is about $45°$, so the graph shows a peak at $45°$.

    b.

    c. It is well known that a projectile has maximum range when the angle is about $45°$ so the graph above shows a peak at about $45°$. This is a good qualitative model.

4. **c.** With only information about the first three miles of the trail, there's no knowing what will happen at mile 5.

5. **c.** Larger rates of change correspond with larger and steeper slopes.

6. **a.** The initial value corresponds with *time* = 0, and this results in an initial *price* = $100.

7. **c.** The coefficient on the independent variable is the slope, which also represents the rate of change. Since this coefficient is negative, we can expect a graph that slopes downward.

8. **a.** The y-intercept $b = 7$, and the slope $m = -2$, and thus the line, in the form $y = mx + b$, has equation $y = -2x + 7$.

9. **b.** A line in the form $y = mx + b$ has slope = m and y-intercept = b, so the slope of both $y = 12x - 3$ and $y = 12x + 3$ is $m = 12$, and they have different y-intercepts (–3 and 3).

10. **c.** Charlie picks $550 - 150 = 400$ apples over the span of two hours, so the slope of the function is $400/2 = 200$.

DOES IT MAKE SENSE?

7. Does not make sense. The graph of a linear function is always a line.

9. Makes sense. The familiar relationship *distance = speed × time* can be solved for *speed* to produce *speed = distance ÷ time*, which is the rate of change in distance with respect to time.

BASIC SKILLS AND CONCEPTS

11. a. Rain depth increases linearly with time.

    b. It appears that the change in depth over the first three hours is four inches, so the slope is 4/3 inches per hour.

    c. The model is realistic if the rainfall rate is a constant 4/3 inches per hour over four hours.

13. a. On a long trip, the distance from home decreases linearly with time.

    b. The distance has decreased by 500 miles over the span of 7 hours, so the slope is $-500 \div 7 = -71.4$ miles per hour.

    c. This is a good model if the speed of travel is a constant 71.4 miles per hour over 7 hours.

15. a. Shoe size increases linearly with the height of the individual.

    b. The change in shoe size was 11 while the change in height was 80 inches, and this produces a slope of $11 \div 80 = 0.1375$ size per inch.

    c. This model is a rough approximation at best, and it would be difficult to find realistic conditions where the relationship held.

17. The *water depth* decreases with respect to *time* at a rate of 2 in/day, so the rate of change is –2 in/day. In 8 days, the water depth decreases by 2 in/day × 8 days = 16 inches. In 15 days, it decreases by 2 in/day × 15 days = 30 in.

19. The *Fahrenheit temperature* increases with respect to the *Celsius temperature* at a rate of 9/5 degrees F per degree C. The rate of change is 9/5 degrees Fahrenheit/degrees C. An increase of 5 degrees C results in an increase of 5 × 9/5 degrees Fahrenheit/degrees C = 9 degrees F. A decrease of 25 degrees C results in a decrease of 25 × 9/5 degrees Fahrenheit/degrees C = 45 degrees F.

21. The *snow depth* increases with respect to *time* at a rate of 3.5 inches per hour. In 6.3 hours, the snow depth will have increased by 3.5 in/hr × 6.3 hr = 22.05 inches. In 9.8 hours, the snow depth will have increased by 3.5 in/hr × 9.8 hr = 34.3 in.

23. The independent variable is time t, measured in years, where $t = 0$ represents today. The dependent variable is price p, measured in dollars. The equation for the price function is $p = 18,000 + 900t$. In 3.5 years, a new car will cost $p = 18,000 + 900 × 3.5 = \$21,150$. This function does not give a good model for car prices.

25. The independent variable is snow depth d, measured in inches, and the dependent variable is maximum speed s, measured in miles per hour. The equation for the speed function is $s = 40 - 1.1d$. To find the depth at which the plow will not be able to move, set $s = 0$ and solve for d.

$$0 = 40 - 1.1d$$
$$1.1d = 40$$
$$d = 40/1.1 = 36 \text{ inches.}$$

The rate at which the speed decreases per inch of snow depth is probably not a constant, so this model is an approximation.

27. The independent variable is time t, measured in minutes, and the dependent variable is rental cost r, measured in dollars. As long as the copy business is willing to prorate rental charges per minute (rather than in 5-minute blocks, even if you do not use the full 5 minutes), the change per minute is $\$2.00/5 = \$0.40$ per minute. This implies the rental cost function is $r = 10 + 0.40t$. To find out how many minutes can be rented for \$25, set $r = 25$, and solve for t.

$$25 = 10 + 0.40t$$
$$15 = 0.40t$$
$$t = 15/0.40 = 37.5 \text{ minutes.}$$

This function gives a very good model of rental costs, provided all of the costs are quoted correctly

29. Let W represent the weight of the dog in pounds, and t the time in years. The slope of the model is $(15 - 2.5) \div (1 - 0) = 12.5$ pounds per year, so the linear function is $W = 12.5t + 2.5$. When the dog is 5 years old, its weight is $W = 12.5 × 5 + 2.5 = 65$ pounds, and at age 10, it weighs $W = 12.5 × 10 + 2.5 = 127.5$ pounds. The model is accurate only for small ages.

31. Let P represent the profit (or loss) realized when selling n raffle tickets. The initial cost of the raffle to the fundraisers is $350 (a negative profit), and the rate of change of the profit with respect to the number of tickets sold is $10 per ticket. Thus the profit function is $P = 10n - 350$. One can see by inspection that 35 tickets must be sold to break even.

33. Let V represent the value of the washing machine, and let t represent time (in years). The value function is $V = -75t + 1200$. Set $V = 0$ and solve for t to find out how long it will take for the value to reach $0.

$$0 = -75t + 1200$$
$$75t = 1200$$
$$t = 1200/75 = 16 \text{ years.}$$

FURTHER APPLICATIONS

35. The slope is $m = 2$, and the y-intercept is $(0, 6)$.

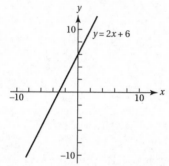

37. The slope is $m = -5$, and the y-intercept is $(0, -5)$.

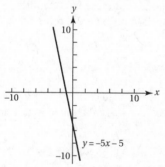

39. The slope is $m = 3$, and the y-intercept is $(0, -6)$.

41. The slope is $m = -1$, and the y-intercept is $(0, 4)$.

Graphs for Exercises 39 and 41:

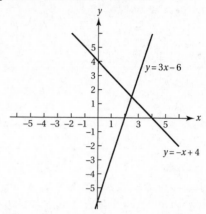

43. The independent and dependent variables are *time* and *elevation*, respectively.

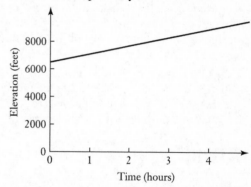

After 3.5 hours, the elevation of the climbers is $6500 + 600 \times 3.5 = 8600$ feet. The model is reasonable provided the rate of ascent is nearly constant.

45. The independent and dependent variables are *number of posters* and *cost*, respectively.

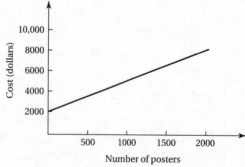

The cost of producing 2000 posters is $2000 + 2000 \times 3 = \$8000$. This function probably gives a fairly realistic estimate of printing costs

47. The independent and dependent variables are *time* and *cost*, respectively.

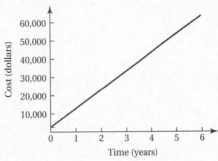

The cost of 6 years of school is 2000 + 10,000 × 6 = $62,000. Provided costs do not change during the six-year period, this function is an accurate model of the cost.

49. a. Half of the fish caught in the second outing are tagged, and this means the proportion of those tagged in the entire population is 1/2. Thus 200/N = 1/2, which implies N = 400 fish.

b. One-fourth of the fish caught in the second outing are tagged, and this means the proportion of those tagged in the entire population is 1/4. Thus 200/N = 1/4, which implies N = 800 fish.

UNIT 9C

QUICK QUIZ

1. **b.** This is the definition of exponential growth.

2. **a.** Q_0 represents the initial value in the exponential model.

3. **b.** The variable r represents the growth rate, expressed as a decimal (and 3% = 0.03).

4. **a.** The population of India at any time t is given by $1{,}270{,}000{,}000 \times 1.013^t$.

5. **c.** Because the value of the dollar is decreasing, r is negative.

6. **c.** If you start out with a positive quantity, and remove half of it, half still remains (and it is a positive quantity, which can never equal 0).

7. **b.** If you know the initial amount and half-life of any exponentially decreasing quantity, you can use $Q = Q_0 \times \left(\frac{1}{2}\right)^{t/T_{\text{half}}}$ to predict the future amount after any time t.

8. **c.** You need to know both the initial amount Q_0 and the amount at some future time t in order to solve the exponential model $Q = Q_0 \times \left(\frac{1}{2}\right)^{t/T_{\text{half}}}$

c. If p represents the proportion of those caught in the second outing that are tagged, we have 200/N = p, which implies $N = f(p) \approx 200/p$.

d. Graph not supplied, though it looks similar to $y = 1/x$, with domain of $0 < p \le 1$. When p is small, N is very large, and when p is close to 1, N is close to 200. Note that the above domain is the domain of the mathematical model, not the set of all possible values of p that could be observed in such a study. Consider a situation where only 200 fish are caught in the first outing from a very large lake with a large fish population. On the second outing, you may find that none of the 200 fish caught is tagged, in which case $p = 0$. This makes sense in the field, but it does not make sense in the mathematical model, because we cannot predict the fish population by using N = 200/0 (division by 0 is not allowed).

e. If p = 15% = 0.15, N = 200/0.15 ≈ 1333 fish.

for t (you must also know T_{half}, but we are given that).

9. **c.** Because 1/16 of the original uranium is present, we know it has undergone 4 halving periods, and since each of these is 700 million years, the rock is 4 × 700 million = 2.8 billion years old.

10. **a.** We know $Q = Q_0(1+r)^t$, and $Q = Q_0 \times 2^{t/T_{\text{double}}}$, and thus $Q_0(1+r)^t = Q_0 \times 2^{t/T_{\text{double}}}$. Divide both sides by Q_0 to get the result shown in the answer.

DOES IT MAKE SENSE?

7. Does not make sense. After 100 years, the population growing at 2% per year will have increased by a factor of $1.02^{100} = 7.245$, whereas the population growing at 1% per year will have increased by a factor of $1.01^{100} = 2.705$. Under only very special circumstances will the first population grow by twice as many people as the second population. (Let Q_0 represent the initial amount of the population growing by 2%, and R_0 the initial amount of the population growing by

7. (continued)

1%. After 100 years, Q_0 will have grown to 7.245Q_0, and thus will have grown by 7.245Q_0 – Q_0 = 6.245Q_0. In a similar fashion, it can be shown that R_0 will have grown by 1.705R_0. We need 6.245Q_0 = 2 × 1.705R_0 in order to satisfy the condition that the first population will grow by twice as many people as the second population. This implies that the ratio R_0/Q_0 = 6.245/3.41 = 1.83. In other words, whenever the initial amount of the second population is 1.83 times as large as the initial amount of the first population, the first population will grow by twice as many people in 100 years. (Try it with initial populations of 100 and 183.)

9. Makes sense. If we know the half-life of the radioactive material, we can create an exponential function that models the quantity remaining as time passes. In order to use this model to determine ages of bones, we only need to measure the amount of radioactive material present in the bone, and the amount it had originally (the latter is usually surmised by making assumptions about conditions present at the time of death).

BASIC CONCEPTS AND SKILLS

11. $2^x = 128 \Rightarrow \log_{10} 2^x = \log_{10} 128 \Rightarrow x \log_{10} 2 =$

$\log_{10} 128 \Rightarrow x = \dfrac{\log_{10} 128}{\log_{10} 2} = 7$

13. $3^x = 99 \Rightarrow \log_{10} 3^x = \log_{10} 99 \Rightarrow x \log_{10} 3 =$

$\log_{10} 99 \Rightarrow x = \dfrac{\log_{10} 99}{\log_{10} 3} = 4.18$

15. $7^{3x} = 623 \Rightarrow \log_{10} 7^{3x} = \log_{10} 623 \Rightarrow$

$3x \log_{10} 7 = \log_{10} 623 \Rightarrow x = \dfrac{\log_{10} 623}{3 \log_{10} 7} = 1.10$

17. $9^x = 1748 \Rightarrow x = \dfrac{\log_{10} 1748}{\log_{10} 9} = 3.40$.

(See Exercise 13 for a similar process of solution.)

19. $\log_{10} x = 4 \Rightarrow 10^{\log_{10} x} = 10^4 \Rightarrow x = 10,000$

21. $\log_{10} x = 3.5 \Rightarrow x = 10^{3.5} = 3162.28$

(See Exercise 19 for a similar process of solution).

23. $3 \log_{10} x = 4.2 \Rightarrow \log_{10} x = 1.4 \Rightarrow x = 10^{1.4}$

$= 25.12$

25. $\log_{10} (4 + x) = 1.1 \Rightarrow 4 + x = 10^{1.1} \Rightarrow x = 8.59$

27. a. $Q = 60,000 \times (1.025)^t$, where Q is the population of the town, and t is time (measured in years).

b.

Year	Population	Year	Population
0	60,000	6	69,582
1	61,500	7	71,321
2	63,038	8	73,104
3	64,613	9	74,932
4	66,229	10	76,805
5	67,884		

c.

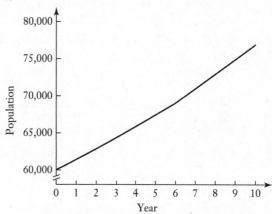

29. a. $Q = 1,000,000 \times (0.93)^t$, where Q is the number of acres of forest left after t years.

b.

Year	Millions of Acres
0	1.00
1	0.93
2	0.86
3	0.80
4	0.75
5	0.70
6	0.65
7	0.60
8	0.56
9	0.52
10	0.48

29. (continued)

c.

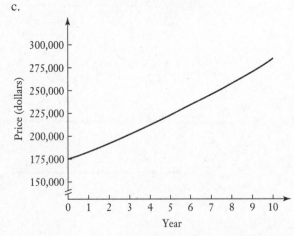

31. a. $Q = 175,000 \times (1.05)^t$, where Q is the average price of a home in dollars, and t is time, measured in years, with $t = 0$ corresponding to 2013.

b.

Year	Average price
0	$175,000
1	$183,750
2	$192,938
3	$202,584
4	$212,714
5	$223,349
6	$234,517
7	$246,243
8	$258,555
9	$271,482
10	$285,057

c.

33. a. $Q = 2000 \times (1.05)^t$, where Q is your monthly salary in dollars, and t is time, measured in years.

b.

Year	Monthly Salary
0	$2000.00
1	$2100.00
2	$2205.00
3	$2315.25
4	$2431.01
5	$2552.56
6	$2680.19
7	$2814.20
8	$2954.91
9	$3102.66
10	$3257.79

c.

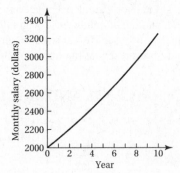

35. Over the span of a year, prices will increase by a factor of $(1.015)^{12} = 1.196$, which is a 19.6% increase.

37. Over the course of a year, prices would have risen by a factor of $\left(1 + \frac{30,000}{100}\right)^{12} = 5.5 \times 10^{29}$, or an annual increase of $5.5 \times 10^{31}\%$. Assuming one day is 1/30 of a month, prices would rise by a factor of $(301)^{(1/30)} = 1.21$ in one day, which is a 21% increase.

39. An exponential model for the animal population is $Q = 1500 \times (0.92)^t$. We need to find the time t that will produce a population $Q = 30$ animals, which amounts to solving $30 = 1500 \times (0.92)^t$ for t.

$$\frac{30}{1500} = 0.92^t \Rightarrow 0.02 = 0.92^t \Rightarrow \log_{10} 0.02 =$$

$$\log_{10} 0.92^t \Rightarrow \log_{10} 0.02 = t \log_{10} 0.92 \Rightarrow$$

39. (continued)

$$t = \frac{\log_{10} 0.02}{\log_{10} 0.92} = 46.92 \text{ years.}$$

41. a. An exponential model for the amount of Valium in the bloodstream is $Q = 50 \times \left(\frac{1}{2}\right)^{t/36}$. After 12 hours, there will be $Q = 50 \times \left(\frac{1}{2}\right)^{12/36} = 39.7$ mg.

b. Ten percent of the initial amount is 5 mg, so solve $5 = 50 \times \left(\frac{1}{2}\right)^{t/36}$ for t:

$$0.1 = (0.5)^{t/36} \Rightarrow \log_{10} 0.1 = \log_{10}(0.5)^{t/36} \Rightarrow$$

$$\log_{10} 0.1 = \frac{t}{36}\log_{10} 0.5 \Rightarrow t = \frac{36\log_{10} 0.1}{\log_{10} 0.5} \Rightarrow$$

$$t = 120 \text{ hours.}$$

43. a. An exponential model for the amount of uranium left is $Q = Q_0 \times \left(\frac{1}{2}\right)^{t/4.5}$, where t is measured in billions of years. We want to determine t when the amount Q is equal to 65% of the original amount, i.e. when $Q = 0.65Q_0$. Solve $0.65Q_0 = Q_0 \times \left(\frac{1}{2}\right)^{t/4.5}$ for t. Begin by dividing both sides by Q_0:

$$0.65 = \left(\frac{1}{2}\right)^{t/4.5} \Rightarrow \log_{10} 0.65 = \log_{10}(0.5)^{t/4.5} \Rightarrow$$

$$\log_{10} 0.65 = \frac{t}{4.5}\log_{10} 0.5 \Rightarrow t = \frac{4.5\log_{10} 0.65}{\log_{10} 0.5} \Rightarrow$$

$$t = 2.8 \text{ billion years old.}$$

b. Following the same process shown in part a, we have $t = \frac{4.5\log_{10} 0.45}{\log_{10} 0.5} = 5.2$ billion years old.

FURTHER APPLICATIONS

45. An exponential model for the amount of radioactive substance is $Q = 3 \times \left(\frac{1}{2}\right)^{t/20}$, where Q is the density of the substance in mg/cm^2, and t is time in years, with $t = 0$ corresponding to the present. This model is valid for times in the past – we need only assign negative values for t. Thus 55 years ago corresponds with $t = -55$, so the density at that time was $Q = 3 \times \left(\frac{1}{2}\right)^{-55/20} = 20.2$ mg/cm^2. One can also solve $3 = Q_0 \times \left(\frac{1}{2}\right)^{55/20}$ for Q_0 to produce the same result.

47. The cart of groceries cost $Q = \$150 \times (1.032)^5 = \$175.59$.

49. a. Experiment with various values of r in $380 = 290 \times (1+r)^{150}$ to find that $r \approx 0.0018$ or 0.18% per year. One can also solve the equation using algebra:

$$380 = 290 \times (1+r)^{150}$$

$$\frac{380}{290} = (1+r)^{150}.$$

Now raise both sides to the reciprocal power of 150, i.e. to the 1/150 power, and then subtract 1 to find $r = \left(\frac{380}{290}\right)^{1/150} - 1 = 0.0018$ or 0.18%.

b. We want to determine t when the amount Q is double the original amount, i.e. when $Q = 2Q_0$. Since $Q_0 = 290$, $2Q_0 = 580$. Solve $580 = 290 \times (1.0018)^t$ for t. Begin by dividing both sides by 290: $2 = (1.0018)^t \Rightarrow$

$$2 = (1.0018)^t \Rightarrow \log_{10} 2 = \log_{10}(1.0018)^t \Rightarrow$$

$$\log_{10} 2 = t\log_{10} 1.0018 \Rightarrow t = \frac{\log_{10} 2}{\log_{10} 1.0018} \Rightarrow$$

$t = 385$ years. The CO_2 concentration will be double its 1860 level in 2245.

c. Answers will vary.

UNIT 10A

QUICK QUIZ

1. **a**. As long as the points are distinct, two points are sufficient to determine a unique line.

2. **b**. The surface of a wall is similar to a plane, which is two-dimensional.

3. **a**. By definition, an acute angle has measure less than 90°.

4. **c**. By definition, a regular polygon is a many-sided figure where each side has the same length, and each internal angle is the same.

5. **b**. By definition, a right triangle has one 90° angle.

6. **a**. The formula for the circumference of a circle of radius r is $C = 2\pi r$.

7. **c**. The formula for the volume of a sphere of radius r is $V = \frac{4}{3}\pi r^3$.

8. **c**. Suppose the side length of a square is denoted by s. Its area is then s^2. If the lengths of the sides are doubled to $2s$, the area becomes $(2s)^2 = 4s^2$, and this is 4 times as large as the original area.

9. **c**. The volume of a sphere whose radius has tripled from r to $3r$ is $V = \frac{4}{3}\pi(3r)^3 = 3^3 \cdot \frac{4}{3}\pi r^3$.

10. **a**. The volume of a block does not increase when it is cut into pieces, though its surface area does, which implies is surface-area-to-volume ratio does as well.

DOES IT MAKE SENSE?

11. Does not make sense. If the highways are, in fact, straight lines, they will not intersect more than once.

13. Makes sense. A rectangular prism is just a box.

15. Does not make sense. Basketballs are in the shape of a sphere (a can is shaped like a right circular cylinder).

BASIC SKILLS AND CONCEPTS

17. One-half of a circle is 180°.

19. One-fifth of a circle is 72°.

21. Two-ninths of a circle is 80°.

23. A 45° angle subtends 45°/360° = 1/8 circle.

25. A 120° angle subtends 120°/360° = 1/3 circle.

27. A 36° angle subtends 36°/360° = 1/10 circle.

29. A 300° angle subtends 300°/360° = 5/6 circle.

31. $C = 2\pi \times 8 = 16\pi$ m = 50.3 m. $A = \pi \cdot 8^2 = 64\pi = 201.1$ m^2

33. $C = 2\pi \times 23/2 = 23\pi$ ft = 72.3 ft. $A = \pi \cdot \left(\frac{23}{2}\right)^2 = 132.25\pi = 415.5$ ft^2

35. $C = 2\pi \times 20 = 40\pi$ mm = 125.7 mm. $A = \pi \cdot 20^2 = 400\pi = 1256.6$ mm^2

37. The perimeter is 4 × 6 = 24 mi. The area is 6 × 6 = 36 mi^2.

39. The perimeter is 2 × 8 + 2 × 30 = 76 ft. The area is 30 × 4 = 120 ft^2.

41. The perimeter is 2 × 2.2 + 2 × 2.0 = 8.4 cm. The area is 2.2 × 2.0 = 4.4 cm^2.

43. The perimeter is 6 + 8 + 10 = 24 units. The area is $\frac{1}{2} \times 8 \times 6 = 24$ units^2.

45. The perimeter is 8 + 5 + 5 = 18 units. The area is $\frac{1}{2} \times 8 \times 3 = 12$ units^2.

47. The two semicircular caps together make a single circle, whose radius is 3 feet. The perimeter, then, is 2 × 8 + 2π × 3 = 34.85 ft, and the area is $6 \times 8 + \pi \times 3^2 = 76.27$ ft^2.

49. The area is $\frac{1}{2} \times 12 \times 11 = 66$ ft^2.

51. The area of the parallelogram is $180 \times 150 = 27{,}000$ yd^2.

53. The pool holds $50 \times 30 \times 2.5 = 3750$ m^3.

55. The duct is in the shape of a cylinder, and so its volume is $\pi \times 1.5^2 \times 40 = 282.7$ ft^3 (note that 18 inches is 1.5 feet). The surface area of the duct is $2\pi \times 1.5 \times 40 = 377.0$ ft^2. We are assuming here that there are no circular end pieces on the duct that need to be painted. If there are end pieces, the area would increase by $2\pi \times 1.5^2 = 14.1$ ft^2.

57. The circumference is greater because it is π diameters (of a tennis ball) long, whereas the height is only 3 diameters tall.

59. The volume of the reservoir is $250 \times 60 \times 12 = 180{,}000$ m^3. 30% of this volume, which is 54,000 m^3, must be added.

61. The first tree has greater volume ($\pi \times 2.1^2 \times 40 = 554.2$ ft^3) than the second tree (whose volume is $\pi \times 2.4^2 \times 30 = 542.9$ ft^3).

63. The surface area of the concert hall will be $30^2 = 900$ times as great as the surface area of the model (because areas always scale with the square of the scale factor).

65. The height of the office complex will be 80 times as large as the height of the model.

67. You would need $80^3 = 512,000$ times as many marbles to fill the office complex (volumes always scale with the cube of the scale factor).

69. Your waist size is a measurement of length, so it has increased by a factor of 4.

71. Your weight is a function of your volume, so it has increased by a factor of 64.

73. Answers will vary. Example: If your waist size is 32 inches, Sam's is $1.2 \times 32 = 38.4$ inches.

75. Think of a soup can (a squirrel) and a trash can (a human), compute their surface areas and volumes, find their surface-area-to-volume ratios, and you will discover that the surface-area-to-volume ratio for squirrels is larger than that of humans. Another way to understand this fact: surface area scales with the square of the scale factor, whereas volume scales with the cube of the scale factor. Thus the surface-area-to-volume ratio decreases when objects are scaled up (and in fact, it scales with the reciprocal of the scale factor).

77. The moon's surface-area-to-volume ratio is four (i.e. the scale factor) times as large as Earth's, (see discussion at the end of Exercise 75).

79. If you understood the comment at the end of Exercise 75, the surface-area-to-volume ratio of the bowling ball should be 1/3 the ratio of the softball, because the bowling ball is 3 times as large. Computing directly: The surface area of the softball is $4\pi \times 2^2 = 50.3$ in^2, its volume is $\frac{4}{3}\pi \times 2^3 = 33.5$ in^3, and its surface-area-to-volume ratio is 1.5. The surface area of the bowling ball is $4\pi \times 6^2 = 452.4$ in^2, its volume is $\frac{4}{3}\pi \times 6^3 = 904.8$ in^3, and its surface-area-to-volume ratio is 0.5. As expected, the ratio of the bowling ball is 1/3 the ratio of the softball.

FURTHER APPLICATIONS

81. a. The book occupies three-dimensional space, so three dimensions are needed to describe it.

   b. The cover of the book occupies a portion of a plane, so two dimensions are needed to describe it.

   c. An edge of the book lies along a line, so a single dimension is adequate for its description.

   d. A point is dimensionless, so anything on the book that corresponds to a point (such as a corner of a page) is of zero dimension.

83. The third line is necessarily perpendicular to both lines due to a theorem from geometry (when two parallel lines are cut by any other line, all three lying in the same plane, alternate interior angles are equal – this amounts to saying that if the angle between one of the parallel lines and the third line is 90°, the angle between the other parallel line and the third line is 90°).

85. a. The surface area of an individual sac is $4\pi \times \left(\frac{1}{6}\right)^2 = 0.349$ mm^2, and the total surface area is 300 million times this value, which is 1.05×10^8 mm$^2 = 105$ m^2. The volume of a single sac is $\frac{4}{3}\pi \times \left(\frac{1}{6}\right)^3 = 0.0194$ mm^3, so the total volume is 5.82×10^6 mm^3 (multiply the individual volume by 300 million).

   b. We need to solve $5.82 \times 10^6 = \frac{4}{3}\pi \times r^3$ for r to find the radius of such a sphere. Multiply both sides of the equation by $\frac{3}{4\pi}$ and take the cube root to find $r = 112$ mm (rounded). This sphere has surface area of $4\pi \times (111.572)^2 = 1.56 \times 10^5$ mm^2. The surface area of the air sacs is $1.05 \times 10^8 \div 1.56 \times 10^5 = 673$ times as large as the surface area of the hypothetical sphere.

   c. We need to solve $1.05 \times 10^8 = 4\pi \times r^2$ for r to find the radius of such a sphere. Divide both sides by 4π and take the square root to find $r = 2891$ mm, or about 2.9 meters (i.e. more than 9 feet). The human lung has a remarkable design as it is able to take advantage of large surface area despite its relatively small volume.

87. The volume of one tunnel is $\frac{1}{2}\pi \times (4)^2 \times 50,000$

$= 1.26 \times 10^6$ m^3, so the volume of all three is 3.8×10^6 m$^3 = 0.0038$ km^3. Depending on whether the dimensions given are for a rough tunnel, or the structure that a motorist sees driving through it, the amount of earth removed may be considerably more than the answer provided.

UNIT 10B

QUICK QUIZ

1. **c.** Each degree contains 60 minutes.

2. **b.** Each degree contains 60 minutes, and each minute contains 60 seconds.

3. **a.** Lines of latitude run east and west on the globe, parallel to the equator.

4. **b.** That you are at 30°S means you are in the southern hemisphere (and thus not in North America). A check of the globe will reveal these coordinates lie in the south Pacific.

5. **b.** Angular size is a function of the distance from the object: when the distance from an object increases, its angular size decreases.

6. **b.** Imagine a solar eclipse, when the Moon is positioned between the Earth and the Sun, and think of the triangle that stretches from a point on Earth to the edges of the moon. Imagine another triangle that continues beyond the Moon to the edges of the Sun (a simple drawing will help you to visualize the situation). These two triangles are similar, and thus we can write the ratio $\frac{\text{diameter of Sun}}{\text{diameter of Moon}} = \frac{400x}{x}$, where x represents the distance from the Earth to the Moon. This ratio implies that the diameter of the Sun is 400 times the diameter of the Moon. One can also appeal to the angular size formula in the text to get the same result.

7. **a.** A 10% grade means the slope is 1/10, which implies a vertical change of 1 unit for every horizontal change of 10 units.

8. **a.** The Pythagorean theorem gives $x = \sqrt{6^2 + 9^2}$.

9. **a.** The statement in answer **a** is an application of the fact that the ratio of corresponding sides in a pair of similar triangles is equal to the ratio of a different set of corresponding sides.

10. **b.** As explained in Example 9 of the text, a circle is the solution to the problem of enclosing the largest possible area with a fixed perimeter.

DOES IT MAKE SENSE?

9. Does not make sense. Points south of the equator experience summer in December.

11. Makes sense. As long as the speaker can ride a bike, a 7% grade should pose no difficulty.

13. Makes sense. Triangle B is just a scaled up version of Triangle A, so they are similar.

BASIC SKILLS AND CONCEPTS

15. $32.5° = 32° + 0.5° \times \frac{60'}{1°} = 32°30'0''$

17. $12.33° = 12° + 0.33° \times \frac{60'}{1°} = 12°19.8' =$

$12° + 19' + 0.8' \times \frac{60''}{1'} = 12°19'48''$

19. $149.83° = 149° + 0.83° \times \frac{60'}{1°} = 149°49.8' =$

$149° + 49' + 0.8' \times \frac{60''}{1'} = 149°49'48''$

21. $30°10' = 30° + 10' \times \frac{1°}{60'} = 30.17°$

23. $123°10'36'' = 123° + 10' \times \frac{1°}{60'} + 36'' \times \frac{1°}{3600''} = 123.18°$

25. $8°59'10'' = 8° + 59' \times \frac{1°}{60'} + 10'' \times \frac{1°}{3600''} = 8.99°$

27. A full circle is 360°, and each degree is 60', so there are $360° \times \frac{60'}{1°} = 21,600'$ in a circle.

29. Madrid is at $40°$N, $4°$W.

31. The latitude changes from $44°$N to $44°$S. To find the longitude, move $180°$ east from $79°$W to arrive at $101°$E. (Note that after moving $79°$, you'll be at the prime meridian, or $0°$, and you still have $101°$ to go). Thus the point opposite Toronto is $44°$S, $101°$E.

33. Buenos Aires is farther from the North Pole because its latitude (35°S) is further south than the latitude of Capetown (34°S).

35. Buffalo is 17° north of Miami, and each degree is 1/360 of the circumference of the Earth. Since the circumference is about 25,000 mi, Buffalo is $\frac{17}{360} \times 25,000 = 1181$ mi, or about 1200 miles from Miami.

37. A quarter is about an inch in diameter, and 3 yards is 108 inches. Thus its angular size is $1 \text{ in} \times \frac{360°}{2\pi \times 108 \text{ in}} = 0.53°$.

39. The Sun's true diameter (physical size) is $0.5° \times \frac{2\pi \times 150,000,000 \text{ km}}{360°} = 1.31 \times 10^6 \text{ km}$.

41. A roof with a pitch of 1 in 4 has a slope of 1/4, which is steeper than a roof with a slope of 2/10.

43. A railroad with a 3% grade has a slope of 3/100 = 0.03, which is not as steep as a railroad with a slope of 1/25 = 0.04.

45. The slope of a 8 in 12 roof is 8/12. The roof will rise $\frac{8}{12} \times 15 = 10$ feet in a horizontal run of 15 feet.

47. Think of a right triangle with side lengths of 6 and 6 – this produces an isosceles triangle, which implies the non-right angles are both 45°. It is possible to have a 7 in 6 roof: it's just a steep roof that rises 7 feet for every 6 feet of horizontal run.

49. The slope of such a road is 20/150 = 0.133, so its grade is 13.3%.

51. a. Walk 6 blocks east (6/8 mi) and 1 block north (1/5 mi) for a total distance of 19/20 = 0.95 mi.
    b. The Pythagorean theorem gives the straight-line distance as $\sqrt{(6/8)^2 + (1/5)^2} = 0.78$ mi.

53. a. Walk 2 blocks west (2/8 mi) and 3 blocks north (3/5 mi) for a total distance of 17/20 = 0.85 mi.
    b. The Pythagorean theorem gives the straight-line distance as $\sqrt{(2/8)^2 + (3/5)^2} = 0.65$ mi.

55. a. Walk 3 blocks east (3/8 mi) and 3 blocks south (3/5 mi) for a total distance of 39/40 = 0.98 mi.
    b. The Pythagorean theorem gives the straight-line distance as $\sqrt{(3/8)^2 + (3/5)^2} = 0.71$ mi.

57. The height of the triangle in Figure 10.31 is $\sqrt{800^2 - 200^2} = 774.6$ ft, and thus the area of the lot is $\frac{1}{2} \times 200 \times 774.6 = 77,460 \text{ ft}^2$, which is $77,460 \text{ ft}^2 \times \frac{1 \text{ acre}}{43,560 \text{ ft}^2} = 1.78$ acres.

59. The height of the triangle in Figure 10.31 is $\sqrt{3800^2 - 600^2} = 3752.3$ ft, and thus the area of the lot is $\frac{1}{2} \times 600 \times 3752.3 = 1,125,700 \text{ ft}^2$, which is $1,125,700 \text{ ft}^2 \times \frac{1 \text{ acre}}{43,560 \text{ ft}^2} = 25.84$ acres.

61. The larger triangle is just a scaled up version of the smaller triangle (or so it appears), and so they are similar.

63. The triangle on the left appears to be an isosceles triangle, whereas the one on the right does not, which means the triangles don't have the same angle measure, and cannot be similar.

65. 8/10 = x/5, which implies $x = 4$. Use the Pythagorean theorem to show that $y = 3$ (we know the triangles are right because the left triangle satisfies $6^2 + 8^2 = 10^2$).

67. x/60 = 10/40, which implies $x = 15$. 50/60 = y/40, which implies $y = 100/3 = 33.3$.

69. Refer to Figure 10.34, and note that a 12-ft fence that casts a 25-foot shadow is equivalent to a house of height h (set back 60 feet from the property line) that casts an 85-ft shadow. We can write the ratio 12/25 = h/85, which implies $h = 40.8$ ft.

71. As in Exercise 69, we can write 12/30 = h/80, which implies $h = 32$ ft.

73. The radius of a circle with circumference of 50 m is $\frac{50}{2\pi} = 7.96$ m, and so its area is $\pi \times (7.96)^2 = 199 \text{ m}^2$. The side length of a square with perimeter of 50 m is 50/4 = 12.5 m, and so its area is $12.5 \times 12.5 = 156 \text{ m}^2$. The area of the circular region is larger.

75. The radius of a circle with circumference of 150 m is $\dfrac{150}{2\pi} = 23.9$ m, and so its area is $\pi \times (23.9)^2 = 1795$ m^2. The side length of a square with perimeter of 150 m is 150/4 = 37.5 m, and so its area is $37.5 \times 37.5 = 1406.25$ m^2. The area of the circular region is larger.

77. For the first can described, the area of the top and bottom is $2 \times \pi \times 4^2 = 100.53$ in^2, and the area of the side is $2\pi \times 4 \times 5 = 125.66$ in^2. The cost of the can is 100.53 in$^2 \times \dfrac{\$1.00}{\text{in}^2} + 125.66$ in$^2 \times \dfrac{\$0.50}{\text{in}^2} = \$163.36$. In a similar manner, it can be shown that the cost of the second can is 157.08 in$^2 \times \dfrac{\$1.00}{\text{in}^2} + 125.66$ in$^2 \times \dfrac{\$0.50}{\text{in}^2} = \$219.91$.

79. Assuming a rectangular box, the most economical shape is a cube, with side length equal to 2 ft (see Example 10 in the text). Such a box has 6 faces, each with area of 4 ft^2, for a total area of 24 ft^2. At $0.15 per square foot, the box would cost $3.60.

FURTHER APPLICATIONS

81. a. The area of the storage region of the Blu-ray disc, which is the area of the entire Blu-ray disc minus the area of the inner circle (that is not part of the storage region), is $\pi \times 5.9^2 - \pi \times 2.5^2 = 89.7$ cm^2.

   b. The density is $\dfrac{50{,}000 \text{ million bytes}}{89.7 \text{ cm}^2} = 557$ million bytes/cm^2.

   c. The length of the groove is $\dfrac{\pi\left((5.9 \text{ cm})^2 - (2.5 \text{ cm})^2\right)}{0.3 \text{ micrometer}} \times \dfrac{10^6 \text{ micrometer}}{1 \text{ m}} \times \dfrac{1 \text{ m}}{100 \text{ cm}} = 2{,}990{,}796$ cm, which is $2{,}990{,}796 \text{ cm} \times \dfrac{1 \text{ m}}{100 \text{ cm}} \times \dfrac{1 \text{ km}}{1000 \text{ m}} \times \dfrac{1 \text{ mi}}{1.6093 \text{ km}} = 18.6$ mi.

83. The throw goes along the hypotenuse of a right triangle, whose length is $\sqrt{90^2 + 90^2} = 127.3$ ft.

85. For simplicity, assume you can row at a rate of 1.0 mph, and you can bike at a rate of 1.5 mph. (The following calculations would yield the same result for any chosen rowing rate.) Using time = $\dfrac{\text{distance}}{\text{rate}}$, the time it takes to bike along the edges of the reservoir is $\dfrac{1.2 \text{ mi} + 0.9 \text{ mi}}{1.5 \text{ mph}} = 1.4$ hr. The time it takes to row along the hypotenuse is $\dfrac{\sqrt{(1.2 \text{ mi})^2 + (0.9 \text{ mi})^2}}{1.0 \text{ mph}} = 1.5$ hr, so biking is faster.

87. a. The volume of the water in the bed is $8 \times 7 \times 0.75 = 42$ ft^3. Divide this volume by the area of the lower room (80 ft^2) to find that the depth of the water is 0.525 ft.

   b. The weight of the water in the bed is $42 \text{ ft}^3 \times \dfrac{62.4 \text{ lb}}{1 \text{ ft}^3} = 2621$ lb.

89. The perimeter of a corral with dimensions of 10 m by 40 m is 100 m. One could use a square-shaped corral with side length of 20 m to achieve the desired area of 400 square meters; this corral would have a perimeter of 80 m, and would require less fencing.

91. Your project will cost $3 \text{ mi} \times \dfrac{\$500}{1 \text{ mi}} + \sqrt{(2 \text{ mi})^2 + (1 \text{ mi})^2} \times \dfrac{\$1000}{1 \text{ mi}} = \$3736$. Using the plan suggested by your boss will cost $4 \text{ mi} \times \dfrac{\$500}{1 \text{ mi}} + \sqrt{(1 \text{ mi})^2 + (1 \text{ mi})^2} \times \dfrac{\$1000}{1 \text{ mi}} = \$3414$.

93. a. With a lot of work and patience, you might be able to come up with the dimensions of the can that has the lowest surface area (and thus lowest cost): $r = 3.8$ cm and $h = 7.7$ cm. One way to save time is to observe that $355 = \pi r^2 h$, which implies that $h = \dfrac{355}{\pi r^2}$. Insert this expression for h into the surface area formula $2\pi r h + 2\pi r^2$ to find that the surface area can be expressed as a function of r: $A = \dfrac{710}{r} + 2\pi r^2$. Since the cost is directly proportional to the surface area, the object is to find the value of r that yields the smallest surface area A. If you graph A on a calculator, and use the trace function, you'll see that $r = 3.8$ at the low point of the graph.

93. (continued)

  b. The dimensions of the optimal can found above imply the can is as wide (diameter) as it is high. Soda cans aren't built that way, probably for a variety of reasons. (Among them: the can needs to fit into the hand of an average human, and the top and bottom of the can cost more because the aluminum is thicker).

95. a. Note that the radius of the cone will be 6 ft (because the height is 1/3 the radius). The volume is $\frac{1}{3}\pi \times 6^2 \times 2 = 75.4 \text{ ft}^3$.

  b. Again, note that $r = 3h$, and solve $1000 \text{ ft}^3 = \frac{1}{3}\pi \times (3h)^2 \times h = 3\pi \times h^3$ for h to find $h = 4.7$ ft.

  c. Assuming 10,000 grains per cubic inch, there are $75.4 \text{ ft}^3 \times \left(\frac{12 \text{ in}}{1 \text{ ft}}\right)^3 \times \frac{10,000 \text{ grains}}{\text{in}^3} = 1.3$ billion grains. The figure of 10,000 grains per cubic inch is very much dependent upon the size of sand grain assumed, which varies.

97. a. The pyramid's height is 481 yd ÷ 3 = 160.3 yards, which is about 1.6 times the length of a football field (excluding the end zones).

  b. The volume of the pyramid is $\frac{1}{3} \times (756 \text{ ft})^2 \times 481 \text{ ft} = 91,636,272 \text{ ft}^3$, which is $91,636,272 \text{ ft}^3 \times \left(\frac{1 \text{ yd}}{3 \text{ ft}}\right)^3 = 3,393,936 \text{ yd}^3$.

  c. Divide the result in part b by 1.5 to get around 2,263,000 blocks.

  d. $2,263,000 \text{ blocks} \times \frac{2.5 \text{ min}}{\text{block}} \times \frac{1 \text{ hr}}{60 \text{ min}} \times \frac{1 \text{ d}}{12 \text{ hr}} \times \frac{1 \text{ yr}}{365 \text{ d}} = 21.5$ years. This is comparable to the amount of time suggested in historical records, which indicates that Lehner's estimate of 10,000 workers is in error. However, we don't have all the facts – perhaps Lehner began with the assumption that, with a work force of 10,000 laborers, one stone could be placed every 2.5 minutes, or maybe he assumed laborers worked on the project 24 hours per day.

  e. The volume of the Eiffel tower is $\frac{1}{3} \times (120 \text{ ft})^2 \times 980 \text{ ft} = 4,704,000 \text{ ft}^3$, which is about 5% of the pyramid's volume.

UNIT 10C

QUICK QUIZ

1. **b.** As noted in the text, fractals successfully replicate natural forms, especially when random iteration is used.

2. **a.** A coastline can be modeled with fractal curves which have dimension between 1 and 2, and by the very definition of fractal dimension, we see that shortening the length of the ruler increases the number of elements (which, in turn, represents the length of the coastline).

3. **b.** The edge of a leaf (think of a maple leaf) has many of the properties of a fractal, such as self similarity, and a length that increases as one decreases the size of the ruler used to measure it.

4. **c.** Fractals often have the property of self similarity, which means that under greater magnification, the underlying pattern repeats itself.

5. **c.** The definition of the fractal dimension of an object leads to fractional values for the dimension.

6. **c.** Data suggest that most coastlines have a fractal dimension of about 1.25.

7. **a.** The curve labeled L_6 in Figure 10.53 is simply the sixth step (or iteration) of the process required to generate the snowflake curve. The snowflake curve itself is denoted by L_∞ (the end product of performing infinite iterations).

8. **c.** An area bounded by a finite curve (it is understood that by this we mean a closed curve of finite length) cannot possibly be of infinite area. If it were possible, we'd have to throw all of our notions about the areas of bounded plane regions out the door. How could the area of a region contained within a region of a finite area be infinitely large?

9. **a.** The definition of a self-similar fractal is that its patterns repeat themselves under greater magnification.

10. **b.** From previous units, we know that if you take, say, an ice cube, and break it into smaller pieces, the aggregate surface area of the pieces is larger than the surface area of the original cube of ice. The same idea is illustrated by the process of creating the Sierpinski sponge. In the first step, when the four subcubes are removed, it can be shown that the surface area increases by a factor of 4/3. In fact, this happens in every step, and thus the surface area becomes infinitely large as the process continues.

DOES IT MAKE SENSE?

9. Makes sense. The boundary of a rectangle is made of straight lines whose lengths are well defined, and so can be measured with a standard ruler (to a reasonable degree of accuracy).

11. Does not make sense. The boundary of the snowflake island is infinitely long, and though its area is finite, it is not computed with a simple "length times width" formula.

13. Makes sense. Data have been collected that shows many boundaries in nature, such as coastlines and edges of leaves, have fractal dimensions greater than one.

BASIC SKILLS AND CONCEPTS

15. According to the definition of fractal dimension, R = 2 and N = 2, so we have $2 = 2^D$, where D is the fractal dimension. It is evident that $D = 1$ in this case, and thus the object is not a fractal (it behaves like an ordinary geometric object with regard to length measurements).

17. According to the definition of fractal dimension, R = 2 and N = 8, so we have $8 = 2^D$, where D is the fractal dimension. It is evident that $D = 3$ in this case, and thus the object is not a fractal (it behaves like an ordinary geometric object with regard to volume measurements).

19. According to the definition of fractal dimension, R = 2 and N = 6, so we have $6 = 2^D$, where D is the fractal dimension. Solving for D using logarithms produces a value of $D = \dfrac{\log_{10} 6}{\log_{10} 2} = 2.585$, and thus the object is a fractal (it does not behave like an ordinary geometric object with regard to area measurements).

21. According to the definition of fractal dimension, R = 5 and N = 5, so we have $5 = 5^D$, where D is the fractal dimension. It is evident that $D = 1$ in this case, and thus the object is not a fractal (it behaves like an ordinary geometric object with regard to length measurements).

23. According to the definition of fractal dimension, R = 5 and N = 125, so we have $125 = 5^D$, where D is the fractal dimension. It is evident that $D = 3$ in this case, and thus the object is not a fractal (it behaves like an ordinary geometric object with regard to volume measurements).

25. According to the definition of fractal dimension, R = 5 and N = 30, so we have $30 = 5^D$, where D is the fractal dimension. Solving for D using logarithms produces a value of $D = \dfrac{\log_{10} 30}{\log_{10} 5} = 2.113$, and thus the object is a fractal (it does not behave like an ordinary geometric object with regard to area measurements).

27. **a.** Suppose that the original line segment has length of 1 unit. When we measure the length of the first iteration of the quadratic Koch curve with a ruler of length 1 unit (or one element), we find the length is 1 unit, the distance from one endpoint to the other. When we shorten the ruler by a factor of $R = 4$ so that its length is 1/4 unit, the length of the first iteration is $N = 8$ elements (i.e., the length is 8/4 = 2 units). This is true at every stage of the process.

   b. According to the definition of fractal dimension, we have $8 = 4^D$. Solving for D yields $D = \dfrac{\log_{10} 8}{\log_{10} 4} = 1.5$. The length of the quadratic Koch curve is infinite, because at each iteration, the length increases by a factor of 2, and thus its length grows without bound as the process continues.

   c. Consider what happens to the area of the quadratic Koch island as we go from the first stage (a square) to the second stage. Along the upper boundary, the area is increased by a small square drawn above the boundary, but it is decreased by a small square drawn below the boundary. This is true along every edge of the original square, and the net effect is no change in the area. The same is true for every step of the process, so the area of the final island is equal to the area of the original square. The coastline of the island is infinite, as noted above in part b.

FURTHER APPLICATIONS

29. When the ruler is reduced by a factor of $R = 9$, there will be $N = 4$ elements found (each of length 1/9). This gives $4 = 9^D$, which when solved leads to $D = \dfrac{\log_{10} 4}{\log_{10} 9} = 0.631$. (Note that this is the answer you get no matter what stage of the process you choose to analyze). The dimension is less than 1 due to the fact that the end result of the process of constructing the Cantor set is a set of isolated points. Though there are infinitely many points, there aren't "enough" of them to constitute a line segment of measurable length (that is, this fractal object does not behave like an ordinary geometric object with regard to length).

31. a. Note that every time the ruler is decreased in length by a factor of $R = 10$, the number of elements increases by a factor of $N = 20$, which leads to $20 = 10^D$. This implies the fractal dimension is $D = \dfrac{\log_{10} 20}{\log_{10} 10} = 1.301$.

    b. Note that every time the ruler is decreased in length by a factor of $R = 2$, the number of area elements increases by a factor of $N = 3$, which leads to $3 = 2^D$. This implies the fractal dimension is $D = \dfrac{\log_{10} 3}{\log_{10} 2} = 1.585$. The fractal dimension is less than two because the standard notion of surface area for ordinary objects does not carry over to the surface area of fractal objects.

    c. Note that every time the ruler is decreased in length by a factor of $R = 2$, the number of volume elements increases by a factor of $N = 6$, which leads to $6 = 2^D$. This implies the fractal dimension is $D = \dfrac{\log_{10} 6}{\log_{10} 2} = 2.585$. A fractal dimension between 2 and 3 is reasonable because such a rock exhibits the same properties as the Sierpinski sponge – it is somewhat "less" than a solid three-dimensional cube because material has been removed in a fractal pattern.

33. The branching in many natural objects has the same pattern repeated on many different scales. This is the process by which self-similar fractals are generated. Euclidean geometry is not equipped to describe the repetitions of patterns on many scales.

UNIT 11A

QUICK QUIZ

1. **b**. Instruments with strings (guitar, violin, piano), woodwinds with reeds (clarinet, bassoon), and instruments with a column of air in a tube (organ pipe, horn, flute) – all produce their sound with an object that vibrates.

2. **a**. In the case of a string vibrating, a single cycle of vibration corresponds to the string moving to its high point and to its low point, so 100 cps means the string is at its high point (and low point) 100 times each second.

3. **a**. Higher pitches go hand in hand with higher frequencies.

4. **c**. The fundamental frequency of a string occurs when the string vibrates up and down along its entire length, and this produces the lowest possible pitch from that string. (Though the text does not discuss it, you may know that a longer wavelength produces a lower frequency – the wavelength can't get any longer for a particular string than the wave associated with the fundamental frequency).

5. **a**. Every time you raise the pitch of a sound by an octave, the pitch doubles.

6. **c**. The 12-tone scale uses twelve equally spaced (in the sense described below) notes for every octave. The factor f by which the frequency changes in moving from one note to the next is constant, which means that in multiplying an initial frequency by f 12 times in a row, you've moved an octave up the scale, and doubled the frequency. This gives us the relationship $f^{12} = 2$, which in turn implies that $f = \sqrt[12]{2}$.

7. **c**. Table 11.1 shows the frequencies of each of the notes in a particular octave; the names of the notes repeat every octave. As you can see in the table, increasing the frequency of middle C by a factor of 1.5 moves you to a G, so if you increase the frequency of the next higher C by a factor of 1.5, you'll end up at the next higher G.

8. **b**. The frequency of each note in the 12-tone scale is related to the note one half-step below it by a multiplicative factor of $f = \sqrt[12]{2} = 1.05946$. This means the frequencies increase by about 5.9% every half-step, which is exponential growth.

9. **b**. The mathematician Fourier discovered that musical sounds are the sum of several constant-frequency waves.

10. **a**. The process of digitizing analog sound changes wave forms that represent sound waves into lists of numbers that can be stored on CDs, and reinterpreted by a CD player to produce a musical sound.

DOES IT MAKE SENSE?

7. Does not make sense. The pitch of a string is a function of the length of the string, not the number of times the string is plucked.

9. Makes sense. The frequency of each note in the 12-tone scale increases by a factor of $f = \sqrt[12]{2} = 1.05946$, and this is exponential growth.

11. Does not make sense. The scratch on the record would be difficult to repair. However, if it is a minor scratch, one might be able to digitize the sound the record produces, and at that point a digital filter could be used to reduce or remove the effects of the scratch.

BASIC SKILLS AND CONCEPTS

13. The next lower octave corresponds to a halving in frequency, so the requested frequencies are 880 cps, 440 cps, 220 cps, and 110 cps.

15.

Note	Frequency (cps)
G	390
G#	413
A	438
A#	464
B	491
C	521
C#	552
D	584
D#	619
E	656
F	695
F#	736

15. (continued)

The table can be generated beginning with the initial frequency of 390 cps, and multiplying by $f = \sqrt[12]{2}$ to produce each successive note. Realize that you will get slightly different values for some of the entries in the table if you generate it from middle C, whose frequency is 260 cps. This is due to rounding errors, and it explains the different values shown in this table and in Table 11.1 at the A, A#, and C entries. We began with an initial value of 390 cps to generate this table, but that's a rounded value for the true frequency of middle G. If we had used an exact value for middle G, we should find that the frequency of the C in this table, which is one octave higher than middle C, is 520 (because frequencies double every octave).

17. a. $260 \times \left(\sqrt[12]{2}\right)^7 = 390$ cps

b. $260 \times \left(\sqrt[12]{2}\right)^9 = 437$ cps

c. An octave is 12 half-steps, so use a total of 19 half-steps. $260 \times \left(\sqrt[12]{2}\right)^{19} = 779$ cps.

d. $260 \times \left(\sqrt[12]{2}\right)^{25} = 1102$ cps

e. An octave is 12 half-steps, so use a total of 39 half-steps. $260 \times \left(\sqrt[12]{2}\right)^{39} = 2474$ cps.

19. To compute the frequency for a note one half-step below a particular frequency, we simply divide by $f = \sqrt[12]{2}$, which is equivalent to multiplying by the factor f raised to the power of –1. Thus to find the note 7 half steps below middle A, use $437 \times \left(\sqrt[12]{2}\right)^{-7} = 292$ cps. Ten half-steps below middle A is $437 \times \left(\sqrt[12]{2}\right)^{-10} = 245$ cps.

FURTHER APPLICATIONS

21. The factor by which the tone has increased is $\left(\sqrt[12]{2}\right)^5 = 2^{5/12} = 1.335$. Because $\left(2^{5/12}\right)^{12} = 2^5$, you need 12 fourths. This is the first time the factor of $2^{5/12}$ turns into a factor of 2, and that is what is necessary to return to the same-named note (raising an octave is a doubling in frequency). There are 5 octaves in the circle of fourths, as evidenced by the factor by which the initial frequency has changed: 2^5.

UNIT 11B

QUICK QUIZ

1. **b.** The principal vanishing point in a painting is the point of intersection of those lines in the painting that are parallel in the real scene, and perpendicular to the canvas. We are assuming here that the tracks are, in fact, perpendicular to the canvas, and that they are straight.

2. **b.** The *horizon line* is the line (horizontal, of course) through the principal vanishing point. According to the principals of perspective, all sets of parallel lines in the real scene intersect at some point along the horizon line in the painting.

3. **a.** Not only does the painting show this fact, but the parallel beams in the ceiling are perpendicular to the canvas, so they should intersect at the principal vanishing point.

4. **c.** There is a vertical line of symmetry in the center of da Vinci's sketch, as both sides of the drawing are nearly identical on either side of this line.

5. **c.** If the letter **W** were reflected across this line, it would appear unchanged from the original.

6. **b.** Rotate the letter **Z** through 180°, and it remains unchanged in appearance.

7. **a.** Because a circle can be rotated through any angle and remain unchanged, it has rotation symmetry. It also has reflection symmetry over any line through its center, for the same reason.

8. **c.** The only regular polygons that admit complete tilings are equilateral triangles, squares, and regular hexagons.

9. **a.** Refer to Figure 11.23 in the text.

10. **c.** A periodic tiling is one where a pattern repeats itself throughout the tiling.

DOES IT MAKE SENSE?

9. Does not make sense. The principal vanishing point is often very apparent in paintings that are rendered in proper perspective (see Figures 11.6 and 11.8, for example).

11. Makes sense. It is the science and art of perspective painting that allows talented artists to render realistic three-dimensional scenes on a two-dimensional canvas.

13. Does not make sense. Only equilateral triangles, squares, and regular hexagons allow for complete tilings. Susan should rent a tile saw.

BASIC SKILLS AND CONCEPTS

15. a. The only obvious vanishing point is depicted in the diagram below. It is not the principal vanishing point because the lines of the road do not meet the canvas at right angles. However, it does lie on the horizon line.

    b.

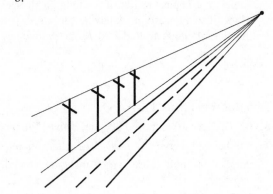

17.

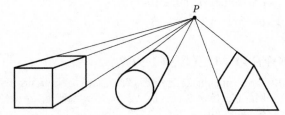

19. a. In order to draw the equally spaced poles, first draw a perspective line along the bases of the existing poles, and draw one along the tops. These two lines should intersect at a vanishing point (directly on, or very near to the horizon). Now measure 2.0 cm along the base line, and draw a vertical pole to meet the top perspective line. Repeat this procedure for the second pole.

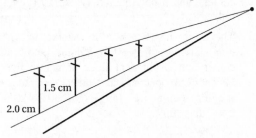

    b. The heights of the poles can be measured directly with a ruler. You should find the first pole you drew is about 1.0 cm, and the second about 0.5 cm. Notice that the lengths of the poles are decreasing in a linear fashion (2.0, 1.5, 1.0, and 0.5 cm). This is due to the fact that the poles are sandwiched between two lines of constant slope, and that the distances between them are equal. (Also note that the diagram in part **a** is not to scale).

    c. If your drawing is an accurate depiction of the scene, the poles are not equally spaced in the actual scene. This is due to perspective: equal distances on the canvas do not correspond with equal distances in the real scene. Because the heights of the poles are assumed to be equal, the short pole you drew in the background is much farther away from you than the first pole you drew, despite the fact that they are equally spaced in the drawing.

21. a. The letters with right/left reflection symmetry are A, H, I, M, O, T, U, V, W, X, and Y.

    b. The letters with top/bottom reflection symmetry are B, C, D, E, H, I, K, O, and X.

    c. The letters with both of these symmetries are H, I, O, and X.

    d. The letters with rotational symmetry are H, I, N, O, S, X, and Z.

23. a. Rotating the triangle about its center 120° or 240° will produce an identical figure.

    b. Rotating the square about its center 90°, 180°, or 270° will produce an identical figure.

    c. Rotating the pentagon around its center 72°, 144°, 216°, or 288° will produce an identical figure.

23. (continued)

    d. Rotating an *n*-gon about its center 360°/*n* (and multiples of this angle less than 360°) will produce an identical figure. There are *n* − 1 different angles of rotational symmetry for an *n*-gon.

25. The figure has reflection symmetries (about vertical and horizontal lines drawn through its center), and it has rotation symmetry (an angle of 180°).

27. The figure has 6 reflection symmetries (about 3 diagonal lines drawn through the petals and the center and 3 diagonal lines drawn in the space between the petals and the center), and it has rotation symmetries (angles of 60°, 120°, 180°, 240°, and 300°).

29.

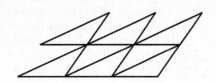

UNIT 11C

QUICK QUIZ

1. **c.** The value of the golden ratio can be derived by solving $\dfrac{L}{1} = \dfrac{L+1}{L}$ for *L*. This equation comes from the idea that the most aesthetically pleasing way to divide a line is one where the ratio of the length of the long piece (*L*) to the length of the short piece (1) is equal to the ratio of the length of the entire line (*L* + 1) to the length of the long piece *L*, and it is this idea that produces the golden ratio.

2. **c.** The fourth number in the Fibonacci series is 3, which is not the value for the golden ratio.

3. **c.** The value of ϕ is about 1.6, and the ratio of the long piece to the short piece should be around 1.6 before claiming the line is divided into the golden ratio. Since 0.6/0.4 = 1.5, these are roughly the correct lengths for a golden ratio.

4. **b.** A golden rectangle is defined to be a rectangle whose ratio of its long side to short side is ϕ.

5. **b.** A golden rectangle is defined to be a rectangle whose ratio of its long side to short side is ϕ. Note than 10 ÷ 6.25 = 1.6.

6. **a.** The golden rectangle became a cornerstone of their philosophy of aesthetics.

31.

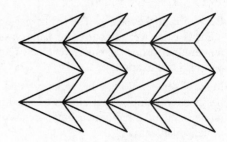

33.

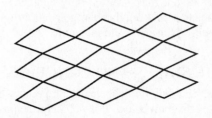

FURTHER APPLICATIONS

35. The angles around a point *P* are precisely the angles that appear inside of a single quadrilateral. Thus, the angles around *P* have a sum of 360°, and the quadrilaterals around *P* fit perfectly together.

7. **a.** Refer to Figure 11.46.

8. **b.** Because the ratio of successive Fibonacci numbers converges to $\phi \approx 1.62$, each Fibonacci number is about 62% larger than the preceding number.

9. **b.** The 21st number in the series is the sum of the preceding two numbers.

10. **b.** Refer to Table 11.3.

DOES IT MAKE SENSE?

7. Does not make sense. The ratio of lengths of Maria's two sticks is 2/2 = 1, which is not the golden ratio.

9. Does not make sense. While circles have a lot of symmetry, they do not embody the golden ratio.

BASIC SKILLS AND CONCEPTS

11. The longer segment should have a length of 3.71 inches, and the shorter segment a length of 2.29 inches.

13. The third and fifth rectangles are golden rectangles.

15. If 5.8 m is the longer side, and S the shorter side, we have $5.8 \div S = \phi$, which means $S = 5.8 \div 1.62 = 3.58$ m. If 5.8 m is the shorter side, and L the longer side, we have $L \div 5.8 = \phi$, which means $L = 5.8\phi = 9.40$ m.

17. If 0.66 cm is the longer side, and S the shorter side, we have $0.66 \div S = \phi$, which means $S = 0.66 \div 1.62 = 0.41$ cm. If 0.66 cm is the shorter side, and L the longer side, we have $L \div 0.66 = \phi$, which means $L = 0.66\phi = 1.07$ cm.

FURTHER APPLICATIONS

19. a. Begin with $\dfrac{L}{1} = \dfrac{L+1}{L}$, and multiply both sides by L to arrive at $L^2 = L+1$, which can be rearranged to produce $L^2 - L - 1 = 0$.

b. The two roots are

$$L = \frac{-(-1) + \sqrt{(-1)^2 - 4(1)(-1)}}{2(1)} = \frac{1+\sqrt{5}}{2} \text{ and}$$

$$L = \frac{-(-1) - \sqrt{(-1)^2 - 4(1)(-1)}}{2(1)} = \frac{1-\sqrt{5}}{2}.$$

The first is $\phi = 1.618034\ldots$.

21. Answers will vary.

23.

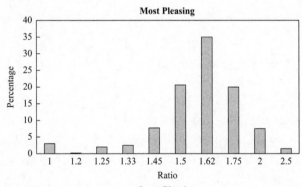

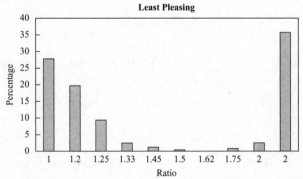

25. a.

a	b	$\dfrac{a+b}{b}$
38	62	1.6129
28	46	1.6087
56	102	1.5490
56	88	1.6364
24	36	1.6667
77	113	1.6814
40	69	1.5797
46	60	1.7667
15	18	1.8333
39	63	1.6190
53	67	1.7910

b. While the values of $\dfrac{a+b}{b}$ are clustered around ϕ, their average is 1.6677 and there are more values greater than ϕ. Therefore, the values of $\dfrac{a+b}{b}$ are not well predicted by ϕ.

c. Answers will vary.

UNIT 12A

QUICK QUIZ

1. **b**. In a three-candidate race, it's possible that none of the candidates receives a majority of the votes.

2. **c**. Gore won 48.38% of the vote, which is not a majority, only a plurality.

3. **b**. A 60% vote is required to end a filibuster, which means 41% of the senators (a minority) can prevent a bill from coming to a vote by staging a filibuster.

4. **a**. In a runoff, the two candidates with the most votes go in a head-to-head competition, and Perot would be eliminated at this stage.

5. **b**. A preference schedule is a ballot where voters list, in order of their preference, the choices before them.

6. **c**. In an election where preference schedules are used, a voter is expected to fill in all choices.

7. **a**. Look at the second row of Table 12.5.

8. **c**. Candidate D showed up in third place on 12 + 4 + 2 = 18 ballots.

9. **b**. Candidate E received only 6 first-place votes.

10. **b**. As shown in the text, a different winner would be declared for each of the five methods discussed (plurality, runoff, sequential runoff, a Borda count, and the Condorcet method).

DOES IT MAKES SENSE?

9. Does not make sense. If both candidates received more than 50% of the vote, the total vote would exceed 100%.

11. Makes sense. Imagine a race with three candidates. Herman could win a plurality of the votes in the first election, with Hanna coming in second, in which case a runoff election would ensue. Hanna would win the election if she beat Herman in the runoff.

13. Makes sense. This is the way most U.S. presidential elections are decided – a candidate wins both the popular and electoral vote.

BASIC SKILLS AND CONCEPTS

15. a. Hayes won $\frac{4,034,142}{8,418,659} = 47.9\%$ of the popular vote while Tilden won $\frac{4,286,808}{8,418,659} = 50.9\%$ of the popular vote. Tilden won a popular majority.

b. Hayes won $\frac{185}{185+184} = 50.1\%$ of the electoral vote while Tilden won $\frac{184}{185+184} = 49.9\%$ of the electoral vote. Hayes won the electoral vote but did not win the popular vote.

17. a. Harrison won $\frac{5,443,633}{11,388,846} = 47.8\%$ of the popular vote while Cleveland won $\frac{5,538,163}{11,388,846} = 48.6\%$ of the popular vote. Neither candidate received a popular majority.

b. Harrison won $\frac{233}{233+168} = 58.1\%$ of the electoral vote while Cleveland won $\frac{168}{233+168} = 41.9\%$ of the electoral vote. Harrison won the electoral vote but did not win the popular vote.

19. a. Clinton won $\frac{44,909,806}{104,423,923} = 43.0\%$ of the popular vote while Bush won $\frac{39,104,550}{104,423,923} = 37.4\%$ of the popular vote. Neither candidate received a popular majority.

b. Clinton won $\frac{370}{370+168} = 68.8\%$ of the electoral vote while Bush won $\frac{168}{370+168} = 31.2\%$ of the electoral vote. Clinton won the popular vote and the electoral vote.

21. a. Bush won $\frac{62,040,610}{122,293,548} = 50.7\%$ of the popular vote while Kerry won $\frac{59,028,439}{122,293,548} = 48.3\%$ of the popular vote. Bush received a popular majority.

b. Bush won $\frac{286}{286+251} = 53.3\%$ of the electoral vote while Kerry won $\frac{251}{286+251} = 46.7\%$ of the electoral vote. Bush won the popular vote and the electoral vote.

124 CHAPTER 12: MATHEMATICS AND POLITICS

23. a. 62/100 = 62% of the senators would vote to end the filibuster, so the filibuster will likely end, and the bill could come to a vote where it would pass.

b. A 2/3 vote of an 11-member jury requires at least 8 votes (because 7/11 = 63.6% < 2/3), so there will be no conviction in this trial. (A hung jury is likely).

c. Only 72% of the states support the amendment (75% is required), so it will fail to pass.

d. It is not likely to become law because 68/100 = 68% of the Senate will support it, and 270/435 = 62% of the House will support it, and this isn't enough for the veto to be overturned.

25. a. Wilson earned 6,286,214/13,896,156 = 45.24% of the popular vote, Roosevelt won 29.69%, and Taft won 25.07%. Wilson won the popular vote by a plurality, but no one won a majority.

b. Wilson won 435/531 = 81.92% of the electoral vote, Roosevelt won 16.57%, and Taft won 1.51%. Wilson won both a plurality and a majority of the electoral vote.

c. If Taft had dropped out of the election, and Roosevelt had won most of Taft's popular votes, then Roosevelt could have won the popular vote. For Roosevelt to win the presidency, his additional popular votes would need to have been distributed among the states in a way that put him ahead of Wilson in many of the states that Wilson had won – Roosevelt would have needed Taft's 8 electoral votes, plus 170 of Wilson's electoral votes to win.

d. If Roosevelt had dropped out of the election and Taft had won most of Roosevelt's popular votes, then Taft could have won the popular vote. However, his additional popular votes would need to have been distributed among states in a way that allowed him to win the electoral vote.

27.

First	A	C	B	C
Second	B	B	C	A
Third	C	A	A	B
	22	20	16	8

29. Use the preference table from Exercise 27.

a. A has 22 votes, B has 16 votes, and C has 28 votes, so C wins.

b. The single runoff would be between A and C, with A having 22 votes and C having 44 votes, so C wins.

c. B is eliminated first, which leaves a runoff between A and C, which is the same as part **b**, so C wins.

d. The count for candidate A is

$22 \times 3 + 20 \times 1 + 16 \times 1 + 8 \times 2 = 118$.

The count for candidate B is

$22 \times 2 + 20 \times 2 + 16 \times 3 + 8 \times 1 = 140$.

The count for candidate C is

$22 \times 1 + 20 \times 3 + 16 \times 2 + 8 \times 3 = 138$.

So B wins.

e. A vs B: B wins 36 to 30.

A vs C: C wins 44 to 22.

B vs C: B wins 38 to 28.

B wins 2 out of 3 pairwise comparisons, so B wins.

31. 22 preferred B to E.

33. With E removed and looking at the first row of the preference table: A would receive 18, B would receive 16, C would receive 12, and D would receive 9.

35. a. A total of 66 votes were cast.

b. Candidate D is the plurality winner (22 first-place votes), but not by a majority.

c. In a runoff, candidate D would win, because 40 voters prefer D to B (who finished in second place in part **b**), while only 26 prefer B to D.

d. In a sequential runoff, the candidate with the fewest votes is eliminated at each stage. Candidate A received the fewest first-place votes in the original results (only 8), so A is eliminated, and those votes are redistributed among the other candidates. Recounting the votes after A has been eliminated, we find that B has 20 first-place votes, C has 16, and D has 30. Thus C is now eliminated, and the votes counted again. In this final stage, we find that B has 26 first-place votes, and D has 40 votes (as in part **b**), so D wins the sequential runoff.

e. In a Borda count, first-place votes receive a value of 4, second-place votes receive a value of 3, and so on. Working through each column of the table, and noting the number of voters who cast such ballots, the count for candidate A is computed as follows:

$20 \times 1 + 15 \times 3 + 10 \times 2 + 8 \times 4 + 7 \times 3 + 6 \times 3 = 156$.

The count for candidate B is

$20 \times 4 + 15 \times 1 + 10 \times 1 + 8 \times 1 + 7 \times 2 + 6 \times 2 = 139$.

The count for candidate C is

$20 \times 2 + 15 \times 2 + 10 \times 4 + 8 \times 2 + 7 \times 1 + 6 \times 4 = 157$.

The count for candidate D is

$20 \times 3 + 15 \times 4 + 10 \times 3 + 8 \times 3 + 7 \times 4 + 6 \times 1 = 208$.

Candidate D wins in this case.

35. (continued)

f. Since there are 4 candidates, there are 6 pairings to consider. The results for each pairing are shown below.

A beats B, 46 to 20.   C beats A, 36 to 30.

D beats A, 52 to 14.   C beats B, 39 to 27.

D beats B, 40 to 26.   D beats C, 50 to 16.

Since D wins more pairwise comparisons than any other candidate, D is the winner.

g. As the winner by all five methods, candidate D is clearly the winner of the election.

37. a. A total of 100 votes were cast.

b. Candidate C is the plurality winner (40 first-place votes), but not by a majority.

c. In a runoff, candidate A would win, because 55 voters prefer A to C.

d. With only three candidates, the sequential runoff method is the same as the standard runoff between the top two candidates, and thus A would win.

e. Using the process illustrated in Exercise 35e, the counts of the three candidates are A = 200, B = 210, and C = 190. Candidate B has the highest count, so B wins a Borda count.

f. Since there are 3 candidates, there are 3 pairings to consider. The results of each pairing are:

B beats A, 55 to 45.   A beats C, 55 to 45.

B beats C, 55 to 45.

Since B wins two of the pairwise comparisons, B is the winner.

g. Candidate A wins by runoff, B wins by a Borda count and pairwise comparisons, and C is the plurality winner. Thus there is no clear winner.

39. a. A total of 90 votes were cast.

b. Candidate E is the plurality winner (40 first-place votes), but not by a majority.

c. Candidate B wins a runoff, because 50 voters prefer B over E, who would receive 40 votes (D is eliminated at the beginning).

d. With only three candidates earning first-place votes, the sequential runoff method is the same as the standard runoff between the top two candidates, and thus B would win.

e. Using the process illustrated in Exercise 35e, the counts of the candidates are A = 260, B = 290, C = 200, D = 290, and E = 310. Candidate E has the highest count, so E wins a Borda cont.

f. Since there are 5 candidates, there are 10 pairings to consider. The results for each pairing are shown below.

A beats B, 60 to 30.   A beats C, 60 to 30.

D beats A, 60 to 30.   E beats A, 70 to 20.

B beats C, 90 to 0.   D beats B, 60 to 30.

B beats E, 50 to 40.   D beats C, 60 to 30.

C beats E, 50 to 40.   E beats D, 70 to 20.

Since D wins more pairwise comparisons than any other candidate (3), D is the winner.

g. Candidates B and E each win two of the five methods, and D wins one method, so all have some claim on the prize. The outcome is debatable.

FURTHER APPLICATIONS

41. a. Fillipo won a plurality, because he won more votes than the other candidates, but no one won a majority, as more than 50% of the vote is required before a majority win can be claimed.

b. Earnest needs just over 23 percentage points to gain a majority and win a runoff. Since 23%/26% = 0.8846, Earnest needs just over 88.46% of Davis's votes.

43. a. King won a plurality because he won more votes than the other candidates, but no one won a majority (King won 382/943 = 40.5% of the vote, which isn't enough).

b. To overtake King, Lord needs 166 additional votes, which is 166/255 = 65.1% of Joker's votes (this would give Lord a total of 472 votes, with King winning only 471 votes).

45. The results of the three head-to-head races are shown below.

A wins over B, 18 to 9.

B wins over C, 19 to 8.

C wins over A, 17 to 10.

No winner can be determined, and the curious situation where voters prefer A to B, B to C, and C to A is called the *Condorcet paradox*, or *voting paradox*.

47. Each voter will record a first, second, third, fourth, fifth and sixth place vote. The points for these places total 6 + 5 + 4 + 3 + 2 + 1 = 21 points. With 30 voters, there are 30 × 21 = 630 total points.

49. a. If enough states have agreed to the compact to guarantee the 270 electoral votes to win the election, the winner of the national popular vote will always have enough electoral votes to win the electoral vote, even if the other states did not award the national popular vote winner any additional electoral votes.

    b. Before the compact can guarantee the 270 votes, the winner of the national popular vote would not be guaranteed enough electoral votes.

    c. Answers will vary.

UNIT 12B

QUICK QUIZ

1. **c**. An election isn't declared fair until all four fairness criteria are satisfied.

2. **a**. Berman wins a plurality because he has the most votes.

3. **b**. Goldsmith won the least number of first-place votes, and so is eliminated in a runoff. Freedman wins Goldsmith's 18 votes, and so beats Berman.

4. **c**. Criterion 1 is only applicable when a candidate wins a majority of votes.

5. **b**. Freedman wins both head-to-head races, and thus by Criterion 2, Freedman should win if this is a fair election.

6. **a**. See Exercise 5.

7. **a**. Candidate X (Freedman) needs to be declared the winner of the first election before Criterion 4 comes into play.

8. **b**. There are plenty of elections where all four fairness criteria are satisfied. Arrow's theorem says only that no election system can be satisfied in *all* circumstances.

9. **b**. The text gives examples of approval voting scenarios that are not fair (and Arrow's theorem guarantees there are such cases).

10. **c**. In a small state, the number of voters per senator is much smaller than in a large state, and thus each voter has more voting power.

DOES IT MAKE SENSE?

5. Makes sense. Karen feels that because she won all head-to-head races, she should win the election so that Criterion 2 is satisfied – otherwise, it's not a fair election.

7. Does not make sense. Assuming the plurality method is used to decide the second election, Table 12.9 claims Criterion 3 is always satisfied (and thus Wendy could not have managed to win).

BASIC SKILLS AND CONCEPTS

9. A candidate who wins a majority is the only person to receive a plurality. Thus the candidate will win the election, and Criterion 1 is satisfied.

11. The following preference schedule is just one possible example.

First	B	A	C	C
Second	A	B	A	B
Third	C	C	B	A
	2	4	2	3

    C is the plurality winner, but A beats both C and B, which means Criterion 2 is violated.

13. Candidate A would win by the plurality method. However if candidate C were to drop out of the election, then B would win by the plurality method, and Criterion 4 would be violated.

15. Assume a candidate receives a majority. In either runoff method, votes are redistributed as candidates are eliminated. But it is impossible for another candidate to accumulate enough votes to overtake a candidate who already has a majority.

17. By the sequential runoff method (which is the same as the top two runoff method with three candidates), candidate B is eliminated, and C wins the runoff. However, candidate B wins head-to-head races against A and C, so Criterion 2 is violated.

19. In the sequential runoff method, candidate B is eliminated first, then candidate A, making C the winner. Now suppose that the 4 voters on the third ballot (ACB) move C up and vote for the ranking CAB. Now A is eliminated first, and B wins the election. Thus criterion 3 is violated.

21. The following preference schedule is just one possible example.

First	A	B	C
Second	B	A	B
Third	C	C	A
	4	3	5

21. (continued)

    After candidate B is eliminated, A wins the runoff. However, if C were to drop out, B would win the election. So Criterion 4 is violated.

23. The following preference schedule is just one possible example.

First	A	B	C
Second	B	C	B
Third	C	A	A
	4	2	1

    Candidate A has a majority of the votes, but loses in a Borda count (A gets 8 points, B gets 9, and C gets 4, using 2, 1, and 0 points).

25. The following preference schedule is just one possible example.

First	A	D	C
Second	B	B	B
Third	C	A	D
Fourth	D	C	A
	4	8	3

    Using the Borda count (with 3, 2, 1, and 0 points), A gets 20 points, B gets 30, C gets 13, and D gets 27 points, making B the winner. However, D wins head-to-head races against all other candidates, and thus Criterion 2 is violated.

27. Using the point system (with 2, 1, and 0 points), candidate A gets 11 points, B gets 11 points, and C gets 14 points, which makes C the winner. However, if Candidate A were to drop from the race, then B would receive 7 points and C would receive 5 points, making B the winner (notice that the point values become 1 and 0 with only two candidates). Thus Criterion 4 is violated.

29. Assume candidate A wins a majority of first-place votes. Then in every head-to-head race with another candidate, A must win (by a majority). Thus, A wins every head-to-head race and is the winner by pairwise comparisons, so Criterion 1 is satisfied.

31. Suppose candidate A wins by the method of pairwise comparisons, and in a second election, moves up above candidate B in at least one ballot. A's position relative to B remains the same or improves, and A's position and B's position relative to the other candidates remains the same, so A must win the second election.

33. The following preference schedule is just one possible example.

First	A	A	B	E	E
Second	B	C	A	B	D
Third	C	D	C	A	B
Fourth	D	E	D	C	A
Fifth	E	B	E	D	C
	1	3	2	1	2

    By the pairwise comparison method, A would beat C, D, and E; B would beat A and C; C would beat D and E; D would beat B and E; and E would beat B. This would lead to A winning. However, if E were to drop out of the election, then A and B would have two pairwise wins, and C and D would have one pairwise win so there is no winner by this method. Criterion 4 is violated because the outcome of the election is changed when E drops out.

35. a. Voting only for their first choices, candidate C wins by plurality with 42% of the vote.

    b. By an approval vote, 28% + 29% = 57% of the voters approve of A, 28% + 29% +1% = 58% approve of B, and 42% of the voters approve of C. The winner is B.

37. The electoral votes per person for each of the states can be computed by dividing the number of electoral votes by the population. The results are as follows:

    New York: 1.48×10^{-6} Illinois: 1.55×10^{-6}

    Rhode Island: 3.81×10^{-6} Alaska: 4.10×10^{-6}

    It is evident that voters in Alaska have more voting power than those in Illinois.

39. The voters in Rhode Island have more voting power than those in Illinois (see Exercise 37).

41. From the greatest to least voting power, per person, the ranking is: Alaska, Rhode Island, Illinois, New York

FURTHER APPLICATIONS

43. A single runoff would eliminate C and D. Candidate A would win the runoff, 23 to 18. There is no majority winner, so Criterion 1 does not apply. Candidate A beats all other candidates one-on-one, and is also declared winner in a single runoff, so Criterion 2 is satisfied. If A is moved up in any of the rankings, it doesn't affect the outcome of the election, so Criterion 3 is satisfied. If any combination of {B, C, D} drops out of the race, the outcome is not changed. So in this case (though not in general), Criterion 4 is satisfied.

45. Candidate A wins the point system (using 3, 2, 1, and 0 points) with 83 points. There is no majority winner, so Criterion 1 does not apply. Candidate A beats all other candidates one-on-one, and is also declared winner by the point system, so Criterion 2 is satisfied. The point system always satisfies Criterion 3. If any combination of {B, C, D} drops out of the race, the outcome is not changed. So in this case (though not in general), Criterion 4 is satisfied.

47. Candidate A wins by a plurality, but not by a majority. There is no majority winner, so Criterion 1 does not apply. Candidate E beats all other candidates one-on-one, but loses by the plurality method, so Criterion 2 is violated. The plurality method always satisfies Criterion 3. If any of B, C, or D were to drop out of the election, the outcome of A winning would change, so Criterion 4 is violated.

49. The candidates E, D, and B are eliminated sequentially, leaving a final runoff between A and C, which C wins. There is no majority winner, so Criterion 1 does not apply. Candidate E beats all other candidates one-on-one, but loses in the sequential runoff method, so Criterion 2 is violated. If candidate C moved up in any of the rankings, the outcome is not affected, so Criterion 3 is satisfied. If A were to drop out of the election, D would win instead of C, so Criterion 4 is violated.

51. We have seen that E beats all other candidates in pairwise races (see Exercise 47) and is the winner by the pairwise comparison method. There is no majority winner, so Criterion 1 does not apply. The pairwise comparison method always satisfies Criteria 2 and 3. It can be shown that if any combination of {A, B, C, D} drops out of the race, the winner is still E, so Criterion 4 is satisfied.

UNIT 12C

QUICK QUIZ

1. **c**. The Constitution does not specify the number of representatives, and the House has, in fact, had 435 members only since 1912. The number briefly rose by two when Hawaii and Alaska were granted statehood (1959), but dropped back to 435 in the next apportionment (and will likely remain at that level, unless a 1941 law is changed).

2. **b**. Apportionment is a process used to divide the available seats among the states.

3. **a**. The standard divisor is defined to be the total U.S. population divided by the number of seats in the House.

4. **b**. The standard quota is defined to be the population of a state divided by the standard divisor. In this case, the standard divisor is 1 million.

5. **b**. In this scenario, the standard divisor would be defined as the population of students divided by the number of teachers, or $25,000 \div 1000 = 25$.

6. **b**. In this scenario, the standard quota would be defined as the population of the school divided by the standard divisor computed in Exercise 5. Thus it would be $220 \div 25 = 8.8$.

7. **c**. Parks Elementary would get the eight teachers because the Hamilton method assigns the extra teacher to the school with the highest standard quota.

8. **c**. The Hamilton, Jefferson, Webster, and Hill-Huntington methods are the only four methods of apportionment that have been used (to date) to assign the seats in the House to various states.

9. **c**. At present, the law states that the Hill-Huntington method is to be used to reapportion seats at every census.

10. **a**. M. L. Balinsky and H. P. Young proved that it is impossible to devise a method of apportionment that satisfies the fairness criteria in all cases.

DOES IT MAKE SENSE?

9. Makes sense. If the number of staff support persons needed in a division depends on the number of employees in that division, an apportionment method would be a good idea.

11. Does not make sense. All apportionment methods have deficiencies, and no single method (of those discussed in the text) is better than the others due to the level of math required to carry it out.

BASIC SKILLS AND CONCEPTS

13. The number of people per representative would be $350,000,000 \div 435 = 804,598$. If the constitutional limit were observed, the number of representatives would be $350,000,000 \div 30,000 = 11,667$.

15. In order to compute the standard quota, the standard divisor must first be computed. The standard divisor is $\dfrac{309 \text{ million}}{435} = 710,344.8$. For Connecticut, the standard quota is $\dfrac{3,574,097}{710,344.8} = 5.03$, which is close to the 5 seats the state actually has, so Connecticut is very slightly underrepresented.

17. The standard divisor is 710,345 (see Exercise 15). For Florida, the standard quota is $\dfrac{18,801,310}{710,344.8} = 26.47$, which is smaller than the 27 seats the state actually has, so Florida is overrepresented.

19. The total number of employees is $250 + 320 + 380 + 400 = 1350$. The standard divisor is then $\dfrac{1350}{35} = 38.57$. For the first division, the standard quota is $\dfrac{250}{38.57} = 6.48$. The standard quotas for the other divisions are 8.30, 9.85, and 10.37, respectively (computed in a similar fashion).

21. The standard divisor is $\dfrac{5000}{100} = 50$

STATE	A	B	C	D	Total
Population	914	1186	2192	708	5000
Standard Quota	$\dfrac{914}{50} = 18.28$	$\dfrac{1186}{50} = 23.72$	$\dfrac{2192}{50} = 43.84$	$\dfrac{708}{50} = 14.16$	100
Minimum Quota	18	23	43	14	98
Fractional Remainder	0.28	0.72	0.84	0.16	N/A
Final Apportionment	18	24	44	14	100

23. Refer to Exercise 19. Using Hamilton's method for the assignments, round each standard quota down to get $6 + 8 + 9 + 10 = 33$ technicians. Since the first and third divisions have the highest remainders, the extra two technicians will be assigned to them, giving a final apportionment of 7, 8, 10, and 10 technicians to each of the four divisions, respectively.

25. The total population is $950 + 670 + 246 = 1866$, so with 100 seats to be apportioned, the standard divisor is $1866 \div 100 = 18.66$. This is used to compute the standard quota and the minimum quotas in the table below. Hamilton's method applied to these three states yields:

State	A	B	C	Total
Pop.	950	670	246	1866
Std. Q.	50.91	35.91	13.18	100
Min. Q.	50	35	13	98
Frac. R.	0.91	0.91	0.18	2
Final A.	51	36	13	100

Assuming 101 delegates, Hamilton's method yields:

State	A	B	C	Total
Pop.	950	670	246	1866
Std. Q.	50.42	35.26	13.32	101
Min. Q.	51	36	13	100
Frac. R.	0.42	0.26	0.32	1
Final A.	52	36	13	101

No state lost seats as a result of the additional available seat, so the Alabama paradox does not occur here.

27. The total population is $770 + 155 + 70 + 673 = 1668$, so with 100 seats to be apportioned, the standard divisor is $1668 \div 100 = 16.68$. This is used to compute the standard quota and the minimum quotas in the table below. Hamilton's method applied to these four states yields:

State	A	B	C	D	Total
Pop.	770	155	70	673	1668
Std. Q.	46.16	9.29	4.20	40.35	100
Min. Q.	46	9	4	40	99
Frac. R.	0.16	0.29	0.20	0.35	1
Final A.	46	9	4	41	100

Assuming 101 delegates, Hamilton's method yields:

State	A	B	C	D	Total
Pop.	770	155	70	673	1668
Std. Q.	46.62	9.39	4.24	40.75	101
Min. Q.	46	9	4	40	99
Frac. R.	0.62	0.39	0.24	0.75	2
Final A.	47	9	4	41	101

No state lost seats as a result of the additional available seat, so the Alabama paradox does not occur here.

29. The total population is $98 + 689 + 212 = 999$, so with 100 seats to be apportioned, the standard divisor is $999 \div 100 = 9.99$. This is used to compute the standard quota and the minimum quotas in the table below. Using a modified divisor of 9.83 instead, we get the modified quotas listed, and the new minimum quotas. Jefferson's method then yields:

State	A	B	C	Total
Pop.	98	689	212	999
Std. Q.	9.81	68.97	21.22	100
Min. Q.	9	68	21	98
Mod. Q.	9.97	70.09	21.57	101.63
N. Min. Q.	9	70	21	100

Since the new minimum quota successfully apportions all 100 seats, we can stop. Note, however, that the quota criterion is violated, because state B's standard quota is 68.97, yet it was given 70 seats.

31. The total population is 979, so with 100 seats to be apportioned, the standard divisor is $979 \div 100 = 9.79$. This is used to compute the standard quota and the minimum quotas in the table below. Using a modified divisor of 9.60 instead, we get the modified quotas listed, and the new minimum quotas. Jefferson's method then yields:

State	A	B	C	D	Total
Pop.	69	680	155	75	979
Std. Q.	7.05	69.46	15.83	7.66	100
Min. Q.	7	69	15	7	98
Mod. Q.	7.19	70.83	16.15	7.81	101.98
NMQ	7	70	16	7	100

Since the new minimum quota successfully apportions all 100 seats, we can stop. The quota criterion is satisfied.

33. After trial and error, a modified divisor of 38.4 was found to work. The results are summarized in the table below.

Division	I	II	III	IV
Number in division	250	320	380	400
Modified Quota	6.51	8.33	9.90	10.42
Number assigned	7	8	10	10

35. An interesting situation arises in this problem: if the standard divisor, standard quotas, and geometric means are computed without rounding, it turns out a modified divisor is not necessary. That is, after computing said values, the Hill-Huntington method immediately produces an apportionment with no leftover technicians, and it is not necessary to seek a modified divisor. The results are summarized in the table below.

Division	I	II	III	IV
Number in division	250	320	380	400
Standard Quota	6.4815	8.30	9.85	10.37
Geometric Mean	$\sqrt{6 \times 7}$ 6.4807	$\sqrt{8 \times 9}$ 8.49	$\sqrt{9 \times 10}$ 9.49	$\sqrt{10 \times 11}$ 10.49
Number assigned	7	8	10	10

Division I is the culprit – the standard quota and geometric mean are so close to one another that it is hard to determine whether one should round up or down (note that the values at the other entries in the table are not shown to the same precision, because the decision about whether to round up or down can be made with less precise values). If the rounded values from Exercise 19 are used, you will need to use a modified divisor in order to find an apportionment that uses all 35 technicians. A divisor of 38.4 is suitable (and realize that modified divisors are not unique, in that divisors close to one another produce the same results).

FURTHER APPLICATIONS

37. The total population is 7710, so with 100 seats to be apportioned, the standard divisor is 7710 ÷ 100 = 77.10. This is used to compute the standard quota and the minimum quotas in the table below. Hamilton's method applied to these three states yields:

State	A	B	C	Total
Pop.	1140	6320	250	7710
Std. Q.	14.79	81.97	3.24	100
Min. Q.	14	81	3	98
Frac. R.	0.79	0.97	0.24	2
Final A.	15	82	3	100

With the addition of a new state D of population 500, for whom 5 new delegates are added, Hamilton's method (with a new standard divisor of 8210 ÷ 105 = 78.19) yields:

State	A	B	C	D	Total
Pop.	1140	6320	250	500	8210
Std. Q.	14.58	80.83	3.20	6.39	105
Min. Q.	14	80	3	6	103
Frac. R.	0.58	0.83	0.20	0.39	2
Final A.	15	81	3	6	105

Even though 5 new seats were added with state D representation in mind, Hamilton's method assigned 6 seats to this state, 1 of them at the expense of state B. Since B lost a seat as a result of the additional seats for the new state, the New State paradox occurs here.

39. a. The total population is 999, so with 100 seats to be apportioned, the standard divisor is 999 ÷ 100 = 99.9. This is used to compute the standard quota and the minimum quotas in the table below. Hamilton's method yields:

State	A	B	C	Total
Pop.	535	334	120	999
Std. Q.	53.55	34.43	12.01	100
Min. Q.	53	34	12	99
Frac. R.	0.55	0.43	0.01	1
Final A.	54	34	12	100

b. Jefferson's method begins as with Hamilton's. As noted in part **a**, the standard divisor is 9.99, so we try lower modified divisors until the apportionment comes out just right. By trial and error, the modified divisor 9.90 is found to work, as documented in the last two rows of the table below. Note that this choice of the modified divisor is not unique, as other nearby values also work.

State	A	B	C	Total
Pop.	535	334	120	999
Std. Q.	53.55	34.43	12.01	100
Min. Q.	53	34	12	98
Mod. Q.	54.04	34.75	12.12	100.91
N. Min. Q.	54	34	12	100

c. Webster's method requires us to find a modified divisor such that the corresponding modified quotas round (not truncate) to numbers that sum to the desired 100 delegates. Inspecting the table in part **a**, we see that the standard divisor and standard quotas are already adequate – we don't need to seek a modified divisor in this case.

State	A	B	C	Total
Pop.	535	334	120	999
Std. Q.	53.55	34.43	12.01	100
Min. Q.	53	34	12	99
Rou. Q.	54	34	12	100

d. The Hill-Huntington method requires us to find a modified divisor such that the corresponding modified quotas, *rounded relative to the geometric mean*, yield numbers which sum to the required 100 delegates. As in part **c**, the standard divisor and standard quotas are already sufficient because they round (per the geometric means) to the desired apportionment. Thus there is no need to seek out a modified divisor in this case.

State	A	B	C	Total
Pop.	535	334	120	999
Std. Q.	53.55	34.43	12.01	100
Min. Q.	53	34	12	99
Geo. M.	53.50	34.50	12.49	-
Rou. Q.	54	34	12	100

The geometric means are for the whole numbers bracketing the standard quotas, namely: $\sqrt{53 \times 54}$ = 53.50, $\sqrt{34 \times 35}$ = 34.50, and $\sqrt{12 \times 13}$ = 12.49. The modified quotas (in this case the standard quotas) are compared to these, and hence 34.43 and 12.01 are rounded down to 34 and 12, respectively, whereas 53.55 is rounded up to 54.

e. All four methods gave the same results.

41. Refer to Exercise 39 for a detailed explanation of the method of solution that yields the following apportionments.

a. Standard divisor = 100.

State	A	B	C	D	Total
Pop.	836	2703	2626	3835	10,000
Std. Q.	8.36	27.03	26.26	38.35	100
Min. Q.	8	27	26	38	99
Frac. R.	0.36	0.03	0.26	0.35	1
Final A.	9	27	26	38	100

b. Modified divisor = 98.3.

State	A	B	C	D	Total
Pop.	836	2703	2626	3835	10,000
Std. Q.	8.36	27.03	26.26	38.35	100
Min. Q.	8	27	26	38	99
Mod. Q.	8.50	27.50	26.71	39.01	101.73
NMQ	8	27	26	39	100

c. Modified divisor = 99.5.

State	A	B	C	D	Total
Pop.	836	2703	2626	3835	10,000
Std. Q.	8.36	27.03	26.26	38.35	100
Min. Q.	8	27	26	38	99
Mod. Q.	8.40	27.17	26.39	38.54	100.50
Rou. Q.	8	27	26	39	100

d. Modified divisor = 99.5.

State	A	B	C	D	Total
Pop.	836	2703	2626	3835	10,000
Std. Q.	8.36	27.03	26.26	38.35	100
Min. Q.	8	27	26	38	99
Mod. Q.	8.40	27.17	26.39	38.54	100.49
Geo. M.	8.49	27.50	26.50	38.50	-
Rou. Q.	8	27	26	39	100

e. The Jefferson, Webster, and Hill-Huntington methods all gave the same result, so one could argue that these yield the best apportionment.

43. a. The total population is 390 students, so with 10 committee positions to be apportioned, the standard divisor is 390 ÷ 10 = 39. This is used to compute the standard quota and the minimum quotas in the table below. Hamilton's method yields:

Group	Soc.	Pol.	Ath.	Total
Pop.	48	97	245	390
Std. Q.	1.23	2.49	6.28	10
Min. Q.	1	2	6	9
Frac. R.	0.23	0.49	0.28	1
Final A.	1	3	6	10

b. Jefferson's method begins as with Hamilton's. As noted in part **a**, the standard divisor is 39, so we try lower modified divisors until the apportionment comes out just right. By trial and error, the modified divisor of 35 is found to work. Note that this choice of the modified divisor is not unique, as other nearby values also work.

Group	Soc.	Pol.	Ath.	Total
Pop.	48	97	245	390
Std. Q.	1.23	2.49	6.28	10
Min. Q.	1	2	6	9
Mod. Q.	1.37	2.77	7.00	11.14
NMQ	1	2	7	10

c. Webster's method requires us to find a modified divisor such that the corresponding modified quotas round (not truncate) to numbers that sum to the desired 10 members. Inspecting the tables in parts **a** and **b**, we see that neither the standard nor modified quotas there work, so we must try other modified divisors. This time, we find that 38 is a suitable modified divisor. Note that this choice of the modified divisor is not unique, as other nearby values also work.

Group	Soc.	Pol.	Ath.	Total
Pop.	48	97	245	390
Std. Q.	1.23	2.49	6.28	10
Min. Q.	1	2	6	9
Mod. Q.	1.26	2.55	6.45	10.26
Rou. Q.	1	3	6	10

43. (continued)

d. The Hill-Huntington method requires us to find a modified divisor such that the corresponding modified quotas, *rounded relative to the geometric mean*, yield numbers that sum to the desired 10 members. The standard divisor of 39 and standard quotas are already sufficient because they round (per the geometric means) to the desired apportionment. Thus there is no need to seek out a modified divisor in this case.

Group	Soc.	Pol.	Ath.	Total
Pop.	48	97	245	390
Std. Q.	1.23	2.49	6.28	10
Min. Q.	1	2	6	9
Geo. M.	1.41	2.45	6.48	-
Rou. Q.	1	3	6	10

The geometric means are for the whole numbers bracketing the standard quotas, namely: $\sqrt{1\times 2}$ = 1.41, $\sqrt{2\times 3}$ = 2.45, and $\sqrt{6\times 7}$ = 6.48. The modified quotas (in this case the standard quotas) are compared to these, and hence 1.23 and 6.28 are rounded down to 1 and 6, respectively, whereas 2.49 is rounded up to 3.

e. The Hamilton, Webster, and Hill-Huntington methods all give the same result, and thus one could argue that each of these yield the best apportionment.

UNIT 12D

QUICK QUIZ

1. **c.** The process of redrawing district boundaries is called redistricting.

2. **b.** This answer does the best job in summarizing the political importance of redistricting.

3. **b.** As shown in Table 12.19 in the text, house elections are decided, on average, by larger margins of victory.

4. **c.** If district boundaries were drawn in a random fashion, one would expect that Republicans would win about 54% of the house seats. Since they won only 6/13 = 46% of the seats, it appears that the

45. Refer to Exercise 43 for a detailed explanation of the method of solution that yields the following apportionments.

a. Note that the total "population" is 2.5 + 7.6 + 3.9 + 5.5 = 19.5 (million dollars).

Standard divisor = 0.78.

Store	Bou.	Den.	Bro.	F. C.	Total
Pop.	2.5	7.6	3.9	5.5	19.5
Std. Q.	3.21	9.74	5.00	7.05	25
Min. Q.	3	9	5	7	24
Frac. R.	0.21	0.74	0.00	0.05	1
Final A.	3	10	5	7	25

b. Modified divisor = 0.76.

Store	Bou.	Den.	Bro.	F. C.	Total
Pop.	2.5	7.6	3.9	5.5	19.5
Std. Q.	3.21	9.74	5.00	7.05	25
Min. Q.	3	9	5	7	24
Mod. Q.	3.29	10.00	5.13	7.24	25.66
NMQ	3	10	5	7	25

c. Modified divisor not necessary.

Store	Bou.	Den.	Bro.	F. C.	Total
Pop.	2.5	7.6	3.9	5.5	19.5
Std. Q.	3.21	9.74	5.00	7.05	25
Min. Q.	3	9	5	7	24
Rou. Q.	3	10	5	7	25

d. Modified divisor not necessary.

Store	Bou.	Den.	Bro.	F. C.	Total
Pop.	2.5	7.6	3.9	5.5	19.5
Std. Q.	3.21	9.74	5.00	7.05	25
Min. Q.	3	9	5	7	24
Geo. M.	3.46	9.49	5.48	7.48	-
Rou. Q.	3	10	5	7	25

e. All four methods give the same results.

district boundaries were set in a way that favored Democrats.

5. **b.** If district boundaries were drawn in a random fashion, one would expect that Democrats would win about 51% of the house seats. Since they won 4/13 = 31% of the seats, it appears that the district boundaries were set in a way that favored Republicans.

6. **c.** Gerrymandering is the drawing of district boundaries so as to serve the political interests of the politicians in charge of the drawing process.

7. **a**. If you concentrate most of the Republicans in a few districts (and essentially concede that they will win in those districts), you will stack the deck in the favor of the Democrats in all the remaining districts.

8. **b**. The courts have generally allowed even very convoluted district boundaries to stand as long as they don't violate existing laws.

9. **a**. If district lines are drawn in such a way as to maximize the number of, say, Democratic seats in the House, they must concentrate the Republican voters in a few districts, and these districts are going to elect Republican representatives (assuming voters vote along party lines).

10. **b**. In an election for a seat that, say, a Democrat is almost guaranteed to win, the real contest occurs in the primary election, rather than the general election. Primaries tend to draw smaller numbers of voters, and those with more clearly partisan interests, and this results in the election of a representative with more extreme partisan views.

DOES IT MAKE SENSE?

7. Makes sense. The district lines in that state are most likely drawn in such a way as to favor Democrats.

9. Does not make sense. If the practice of gerrymandering is alive and well in your state, you probably shouldn't expect that the percentage of voters from the various political parties is accurately reflected in the percentage of representatives from those parties that win elections.

11. Does not make sense. Current laws require that district lines should be drawn to produce contiguous districts.

BASIC SKILLS AND CONCEPTS

13. a. In 2010, the percentage of votes cast for Republicans was $\dfrac{2053}{2053+1611}=56\%$, and for Democrats it was $\dfrac{1611}{2053+1611}=44\%$. In 2012, the percentage of votes cast for Republicans was $\dfrac{2620}{2620+2412}=52\%$, and for Democrats it was $\dfrac{2412}{2620+2412}=48\%$.

b. In 2010, the percentage of House seats won by Republicans was 13/18 = 72%, and for Democrats it was 5/18 = 28%. In 2012, the percentage of House seats won by Republicans was 12/16 = 75%, and it was 4/16 = 25% for Democrats.

c. In both 2010 and 2012, the percentage of House seats won by Republicans was significantly larger than the percentage of votes they received.

d. Redistricting does not seem to have had an effect on the distribution of representatives.

15. a. In 2010, the percentage of votes cast for Republicans was $\dfrac{3058}{3058+1450}=68\%$, and for Democrats it was $\dfrac{1450}{3058+1450}=32\%$. In 2012, the percentage of votes cast for Republicans was $\dfrac{4429}{4429+2950}=60\%$, and for Democrats it was $\dfrac{2950}{4429+2950}=40\%$.

b. In 2010, the percentage of House seats won by Republicans was 23/32 = 72%, and for Democrats it was 9/32 = 28%. In 2012, the percentage of House seats won by Republicans was 24/36 = 67%, and it was 12/36 = 33% for Democrats.

c. In 2010, the percentage of House seats won by Republicans was close to the percentage of votes they received. In 2012, the percentage of House seats won by Republicans was higher than the percentage of votes they received.

d. Because of the change noted in part **c**, it's plausible that redistricting had an effect on the distribution of representatives.

17. a. In 2010, the percentage of votes cast for Republicans was $\dfrac{2034}{2034+1882}=52\%$, and for Democrats it was $\dfrac{1882}{2034+1882}=48\%$. In 2012, the percentage of votes cast for Republicans was $\dfrac{2710}{2710+2794}=49\%$, and for Democrats it was $\dfrac{2794}{2710+2794}=51\%$.

b. In 2010, the percentage of House seats won by Republicans was 12/19 = 63%, and for Democrats it was 7/19 = 37%. In 2012, the percentage of House seats won by Republicans was 13/18 = 72%, and it was 5/18 = 28% for Democrats.

17. (continued)

c. In 2010, the percentage of House seats won by Republicans was higher than the percentage of votes they received. In 2012, the percentage of votes received by Republicans decreased only by a small amount, yet the percentage of seats won by Republicans went up by a significant margin.

d. Because of the change noted in part **c**, it's plausible that redistricting had an effect on the distribution of representatives.

19. a. The most likely distribution would be 8 Republican and 8 Democrat seats.

b. The maximum number of Republican seats that could be won is 15. In order to get a majority of Republican voters in as many districts as possible, you need to load up one district with Democrats. Each district has 625,000 people – imagine drawing district lines so that all 625,000 people were Democrats in a particular district. That district would elect a Democrat for its representative, but the remaining 15 districts could be drawn to have Republican majorities, and they would elect Republicans. (Note: With the assumption that voting is to take place along party lines, all you'd really need to do is move 1 Democrat from each of 15 districts into the remaining district, and then take 15 Republicans from that district and distribute them evenly across the other 15 districts. This would produce a Republican majority in 15 districts. Of course this assumes that each district begins with exactly 312,500 Republicans, and 312,500 Democrats.) Reverse the above logic to convince yourself that the minimum number of Republican representatives is 1.

21. a. The most likely distribution would be 6 Republican and 6 Democrat seats.

b. The maximum number of Republican seats is 11, and the minimum number is 1. (Refer to Exercise 19b; the logic is the same).

23. a. The most likely distribution would be 15 Republican and 0 Democrat seats. This is due to the fact that with random district lines, one would expect an 80% majority in favor of Republicans in every single district, and as long as voting takes place along party lines, all of these districts will elect Republican representatives.

b. From part **a**, we see that the maximum number of Republican seats is 15. In order to find the minimum number, first observe that each district has 500,000 people, and the entire state has 1,500,000 Democrats. A particular district will elect a Democrat only when the district has a Democrat majority. This requires at least 250,001 Democrats (and 249,999 Republicans), with the assumption that voting takes place along party lines (and that everyone votes). However, there are only 1,500,000 Democrats to spread around, and since $1,500,000 \div 250,001 = 5.999976$, the maximum number of districts where one could find a Democrat majority is 5. This implies that the minimum number of Republican seats is 10.

25. There are several solutions. Perhaps the easiest to describe is one where the boundaries divide the state into 8 rectangles, each 4 blocks wide and 2 blocks high.

27. There are several solutions. Perhaps the easiest to describe is one where the boundaries divide the state into 8 rectangles, each 4 blocks wide and 2 blocks high.

29. It is not possible to draw district boundaries that satisfy the conditions given. Three Democrats are needed in a single district for a Democrat majority, and thus a valid solution would require three Democrats in each of three districts, for a total of nine Democrats. But there are only eight Democrats in the state.

31. Answers will vary (there are several solutions).

FURTHER APPLICATIONS

33. Answers will vary for the first three cases (there are several solutions). The last case, where 4 Republicans are elected, is not possible. Five Republicans are needed in a single district for a Republican majority, and thus a valid solution would require five Republicans in each of four districts, for a total of 20 Republicans. But there are only 18 Republicans in the state.

35. Answers will vary.

37. It is not possible for one party to win every House seat for the situation described. To show this for the example provided (20 voters and four districts), note that 3 voters of the same party are required for a majority in a single district. Thus a majority in all four districts would require 12 voters of the same party, which is impossible, because there are only 10 such voters. This argument can be generalized to give a proof of the statement in the exercise.